The Atlas of Changing South Africa

Since the first edition was published in 1994 as *The Atlas of Apartheid* there has been enormous change in South Africa. Gradually apartheid is being dismantled but in many sectors the effects have not yet been reversed. In this new edition A.J. Christopher examines the spatial impact of apartheid during the period of National Government from 1948 to 1994, and the legacy it has left for South Africa at the beginning of the twenty-first century.

Apartheid was basically about the control of space and specific places. Intent upon maintaining white minority rule, despite local and international resistance, the government thought in terms of drawing lines on maps and on the ground to separate the South African peoples into discrete, legally defined groups in a classic example of divide-and-rule. Segregation operated at many levels and on many scales, from 'petty apartheid' exemplified by separate entrances to buildings and residential areas to 'grand apartheid' involving separate nation-states. It is remarkable that those structures associated with petty and grand apartheid have been dismantled very rapidly, but those associated with the ownership and occupation of land have been extremely persistent.

In providing a comprehensive introduction to and detailed analysis of the policy of apartheid and its aftermath through more than 170 maps, *The Atlas of Changing South Africa* makes a unique contribution. By presenting it in visual, spatial forms most relevant to its conception, it illustrates the various levels of operation of the policy and its wider implications, globally as well as nationally.

A.J. Christopher is Professor of Geography at the University of Port Elizabeth, South Africa.

The Atlas of Changing
South Africa

A.J. Christopher

London and New York

First published 1994 as *The Atlas of Apartheid*
by Routledge
11 New Fetter Lane, London EC4P 4EE

Simultaneously published in the USA and Canada
by Routledge
29 West 35th Street, New York, NY 10001

Second edition 2001

Routledge is an imprint of the Taylor & Francis Group

Typeset in Galliard by RefineCatch Limited, Bungay, Suffolk
Printed and bound in Great Britain by
T.J. International Ltd, Padstow, Cornwall

British Library Cataloguing in Publication Data
A catalogue record for this book is available from the British Library

Library of Congress Cataloging in Publication Data
Christopher, A.J.
   The Atlas of Changing South Africa / A.J. Christopher. – 2nd ed.
      p.   ; cm.
   Rev. ed. of: The atlas of apartheid / A.J. Christopher. 1st ed., c1994.
   'Simultaneously published in the USA and Canada by Routledge.'
      – T.p. verso.
   'First published 2001 by Routledge' – T.p. verso.
   Includes bibliographical references and index.
   1. Apartheid – South Africa – Maps. 2. Anti-apartheid movements – South Africa – Maps.
   3. Apartheid – South Africa. 4. Anti-apartheid movements – South Africa. I. Title.
   G2566.E625 C5 2000
   305.8′00968′022–dc21

                                                            00-030049

ISBN 0–415–21177–8 (hbk)
ISBN 0–415–21178–6 (pbk)

Contents

Figures

Preface

The Atlas of Changing South Africa is offered in response to the successful transition from apartheid to democracy since 1990. It seeks to trace the spatial impact of the complex policies that have shaped the country in the past and have been bequeathed as a legacy for the future. The period between 1990 and 1999 was epoch making in South Africa, witnessing the peaceful political transition and the formal repudiation of apartheid. However, apartheid remains the key to an understanding of the multitude of problems facing the country, as it pervaded every aspect of life. Thus, although this work looks forward to the twenty-first century, the heritage of the twentieth cannot be unravelled in a decade. The *Atlas* accordingly constitutes a necessary reminder of the spatial impact of that heritage and then seeks to show why the new South Africa appears as it does and why the struggle for the control of space is vital to an understanding of the country.

The *Atlas* owes much to the cartographic skills of Wilma Grundlingh who drew the maps for the original edition and has drawn the more than forty additional maps for the new edition. The work thus benefits enormously from her expertise and dedication. I also wish to express my deep gratitude to my wife, Anne, for the continued and inestimable support and insights she has offered in the production of both the original and the present editions.

A.J. Christopher
Port Elizabeth, July 1999

Introduction

South Africa has been the scene of a number of momentous social engineering projects from colonialism and segregationism to apartheid and, currently, the democratic transformation. All of these had profound spatial implications and left significant legacies in the geography of the country. Indeed, successive governments have been deeply implicated in the ordering of society within legally defined spaces for the attainment of political objectives. Apartheid, as implemented by the National Party between 1948 and 1994, was especially concerned with the control of space, notably its occupation and use on a racial basis. Since 1994 the African National Congress-led government has directed substantial resources towards the elimination of the inequalities inherited from its predecessor. It is still too early to talk about the 'post-apartheid' era in other than political terms as the spatial systems and physical structures inherited from the past were solidly constructed and are proving to be remarkably resilient. Thus the *Atlas* is organized around the dominant concept of apartheid, before, during and after its active implementation.

The term 'apartheid' was one of the most emotive in the political vocabulary of the second half of the twentieth century. The Afrikaans word *apartheid* has become the universally employed nomenclature for legalized and enforced racial and ethnic discrimination, notably in the fields of residential segregation, job opportunity and political rights. In its original form, derived from the parent Dutch language, the word meant 'separateness' or 'apartness'. However, in the twentieth century it assumed a political usage denoting a legally enforced policy to promote the political, social and cultural separation of racially defined communities for the exclusive benefit of one of these communities. As such, the word entered certain languages, without change, from Portuguese to Lithuanian and was transliterated into some including Russian and Arabic. Furthermore, the term ceased to be nationally specific and its occurrence was widely identified in other countries from the United States to Morocco (Abu-Lughod 1980; Massey and Denton 1993).

However, it was in South Africa that apartheid assumed the full meaning with which it is usually associated. In 1973 the United Nations went so far as to condemn the policy of apartheid as a 'Crime against Humanity', as it was seen to violate the basic human rights of the majority of the population (Coleman 1998). In 1999 the South African National Assembly similarly declared apartheid to be a crime. The task of overcoming the complex impact of this policy, which intruded into every facet of people's lives, remains as the most significant legacy in the development of the country in the new millennium, long after its formal repudiation.

Yet apartheid as theoretically conceived was steeped in Christian Calvinist fundamentalism and early twentieth-century scientific theories of race (Dubow 1995). White Afrikaner racial preservation and the rejection of miscegenation lay at the core of the policy and hence international repudiation stemmed from the repudiation of similar Nazi ideologies in the Second World War. However, in promoting racism, it was possible for one of the more extreme Nationalist ideologues, Professor G.A. Cronje, to write in 1945:

> The racial policy which we as Afrikaners should promote must be directed to the preservation of racial and cultural variety. This is because it is according to the Will of God, and also because with the knowledge at our disposal it can be justified on practical grounds. . . . The more consistently the policy of apartheid could be applied, the greater would be the security for the purity of our blood and the surer our unadulterated European racial survival. . . . Total racial separation . . . is the most consistent application of the Afrikaner idea of racial apartheid.
>
> (Study Commission 1981: 41)

It was only in 1948, when the National Party came to power in South Africa and proceeded to implement the principles of apartheid within a country already deeply immersed in colonial segregationism, that pragmatism gave way to ideology. The consequences were far reaching, with significant transformations of both the urban and the rural areas, large-scale forced removals of people, and the redrawing of the internal political structures of the state. Race became the dominant element in determining the rights, political and legal, of the members of the population. The map of South Africa and its towns and cities were redrawn on racial lines with different rights assigned to different groups of people within the different zones.

The whole process was fuelled by the determination of the politically and economically dominant White group to retain power over the country in the face of rising demands for political rights by the African majority. Unlike other mid-latitude British dominions, the immigrant populations never achieved demographic dominance, as opposed to political and economic domination, in South Africa. Even at its relative height in the first half of the twentieth century the White population only amounted to a fifth of the total for the country as a whole (Figure 0.1). This proportion subsequently declined to stand at approximately 10 per cent at the beginning of the new millennium.

Economic dominance was most clearly illustrated by the disparity in income and spending power between the various sectors of the population (Adam and Giliomee 1979). In 1946 the White one-fifth of the population controlled nearly three-quarters of personal income (Figure 0.2). Only in the 1980s did the White share begin to fall significantly, although the White one-tenth of the population still accounted for over 40 per cent of personal income in 1999 (*Sunday Times*, 14 March 1999). By this stage the disparities in incomes were greater than those experienced in Brazil (*South Africa Survey* 1998).

The enforcement of job reservation, on a racial basis, and hence the maintenance of relatively high White incomes, was emotionally linked to the preservation of continued White rule and 'White civilization'. This was baldly stated by Mr B.J. Schoeman, Minister of Labour, in 1954, in the debate on the Industrial Conciliation Bill:

> this provision is against economic laws. The question, however, is this: What is our first consideration? Is it to maintain the economic laws or is it to ensure the continued existence of the European race in this country? . . . I want to say that if we reach the stage where the Native can climb to the highest rung of our economic ladder and be appointed in a supervisory role over Europeans, then the other equality,

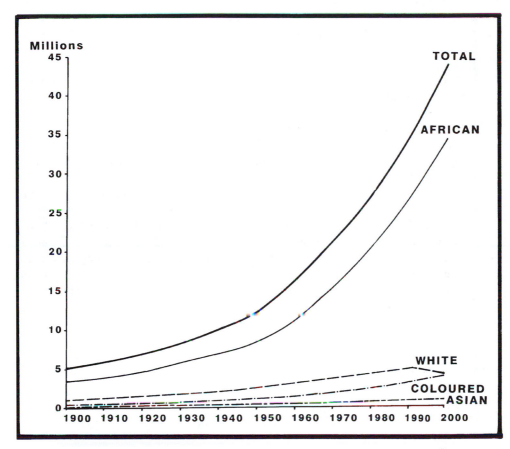

Figure 0.1 Population of South Africa, 1900–2000
Source: Statistics from census reports 1904–96 and projections to 2000

namely political equality, must inevitably follow and that will mean the end of the European race.

(*Hansard* 1954: col. 5854)

The privileged political and economic position of the White population was kept in place through the ruthless enforcement of an elaborate series of laws which were constantly amended and extended to remove any loopholes which might be discovered. Many of the enactments, including the Natives Land Act (1913), the Native Trust and Land Act (1936) and the Natives (Urban Areas) Act (1923), predated the National Party's accession to power, but were rigorously enforced and amended to serve new purposes after 1948. Other legislation was entirely new, notably the Population Registration Act (1950), the Group Areas Act (1950) and the Promotion of Bantu Self-government Act (1959), which sought to establish a more comprehensive and ordered regime. Numerous other enactments determined in detail the activities of the majority of the population, the bureaucracy was enlarged to administer the legislation, and the police force extended to supervise obedience to the law.

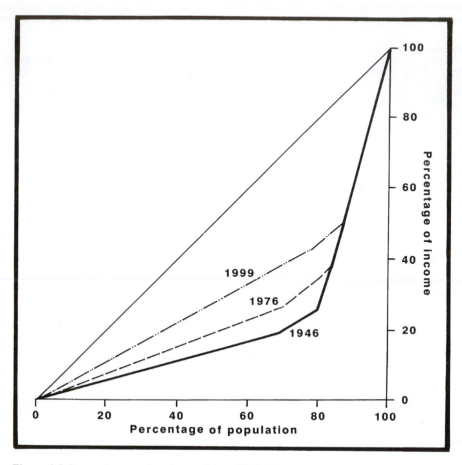

Figure 0.2 Per capita spending inequalities, 1946–99
Source: Statistics from H. Adam and H. Giliomee (1979) *Ethnic Power Mobilized: Can South Africa Change?*, New Haven, CT: Yale University Press, and *South Africa Survey 1997/98*, Johannesburg: South African Institute of Race Relations

The apartheid legislation was initially designed to establish urban racial residential segregation. The Prime Minister, Dr D.F. Malan, was most emphatic that this was the key to apartheid, when debating the Group Areas Bill in 1950:

> The Group Areas Bill . . . is a measure which is proposed by this side of the House to carry out its policy of apartheid, and in that respect it undoubtedly is the most important of all . . . will mean a fresh start for South Africa, because it heralds a new period. I do not think there is any other Bill, affecting the relations between the different races, the non-Europeans and the Europeans in this country, which determines the future of South Africa and of all population groups as much as this Bill does.
>
> (*Hansard* 1950: col. 7722)

Subsequently attention was directed towards the exclusion of anyone not classified as White from the political process of the country. The Population Registration Act thus formed the cornerstone of the entire policy as everyone was required to be classified into a distinct racial group. In such a system there was no room for ambiguity. In 1949 the government had passed the Prohibition of Mixed Marriages Act, with the object, in the words of Dr T.E. Donges, Minister of the Interior: 'to check blood mixture, and as far as possible promote racial purity' (*Hansard* 1949: col. 6164). Not content with a dual White–Black grouping, the government introduced a complex classification which sought to divide the indigenous population from those of mixed parentage and from other immigrant groups. Furthermore, the indigenous population itself was subdivided, according to home or inherited language, into a series of 'national units' or incipient nations.

Separation on the personal level was deemed essential for the survival of the White race. A series of laws was enacted in order to prevent contact between Whites and other people. The Minister of Justice, Mr C.R. Swart, introducing the Reservation of Separate Amenities Bill in 1953, sought to show how such measures were indispensable:

> If a European has to sit next to a non-European at school, if on a railway station they are to use the same waiting-rooms, if they are continually to travel together on the trains and sleep in the same hotels, it is evident that eventually we would have racial admixture, with the result that on the one hand one would no longer find a purely European population and on the other hand a non-European population.
>
> (*Hansard* 1953: col. 1053)

It might be added that amenities were to be separate but not equal:

> it was never the intention of Parliament to say . . . that if you reserve something for one group, equal provision should be made in every respect for the other group. In our country we have civilized people, we have semi-civilized people and we have uncivilized people. The Government of this country gives each section facilities according to the circumstances of each.
>
> (*Hansard* 1953: cols. 1054–5)

Inequality was most marked in the realm of education, where Dr H.F. Verwoerd, when Minister of Native Affairs, piloted the Bantu Education Bill through parliament in 1953:

> Racial relations cannot improve if the wrong type of education is given to Natives. They cannot improve if the result of Native education is the creation of frustrated people who, as a result of the education they received, have expectations in life which circumstances in South Africa do not allow to be fulfilled immediately, when it creates people who are trained for professions not open to them.
>
> (*Hansard* 1953: col. 3576)

Echoes of Nazi education policies for the subjugated peoples of eastern Europe in the early 1940s are evident. Martin Bormann, Hitler's Party Secretary and Deputy Fuehrer, summed up the official attitude succinctly: 'Education is dangerous. It is enough if they can count up to 100. . . . Every educated person is a future enemy' (Shirer 1964: 1118).

The ultimate apartheid experiment was the excision of the African areas from the country, and hence the removal of the African population from the political process. At first the National Party government was ambivalent on the issue, prompting one author to suggest that 'The Myth of the "Grand Design"' had been propagated politically but was not consistent with the record (Posel 1991). It was only in 1959 that the new Prime Minister, Dr H.F. Verwoerd, injected a greater sense of urgency and direction into the debate on the political future of the African population:

> I say that if it is within the power of the Bantu and if the territories in which he now lives can develop to full independence, it will develop that way. Neither he nor I will be able to stop it and none of our successors will be able to stop it.
>
> (*Hansard* 1959: col. 6221)

Such was the success of Dr Verwoerd's partition policy that by the 1970s several African political leaders had accepted the concept of independence from South Africa and were concerned with how much territory they could obtain. President Lucas Mangope of Bophuthatswana argued for a more generous division of South Africa in order to obtain international credibility at his country's independence celebrations in 1977:

> Independence and consolidation are only two sides of the same coin. If any one side of this coin lacks integrity and credibility, the coin will be regarded as faked, and it will be rejected. It is quite self-evident, that the Achilles' heel of Bophuthatswana's credibility is the present state of consolidation, or rather non-consolidation.
>
> (*Eastern Province Herald*, 7 December 1977)

Apartheid in its most basic form involved the removal of anyone not considered to be a White from areas deemed to be for the occupation or enjoyment of the White population. Accordingly, massive forced population movements in both the urban and the rural areas began in the initial years of the National Party's administration. The reorganization of South Africa's towns and cities was a long-term process extending over more than two decades. Similarly, the removal of the 'surplus' African population from the 'White' rural areas was spread over as long a period of time. The lengthy time span allowed for an evolution of the apartheid concept from its basic segregationist origins to the more subtle concepts of 'separate development' and 'separate freedoms', even if the intent of maintaining White supremacy remained the same.

Apartheid was thus conceived as a spatial policy, with markedly geographical consequences. Lines were drawn on maps at various scales, and people were evicted and resettled to fit the lines. The administrators were acutely place conscious, and put a heightened emphasis on the distinctiveness of particular places and communities. Not only were towns redesigned into separate sectors, even with separate administrations, but an entirely new political map of the country unfolded as the policy developed. State partition became the official aim by the 1970s, with South Africa fragmented into a series of African nation-states and a large White-controlled rump entity. Policies were pursued to increase the political and economic viability of the proposed African nation-states, but without weakening the position of the White state. Industrial development, transport planning and regional planning were all undertaken with the goal of creating a new, smaller, but Whiter, South Africa.

Problems multiplied as White minority rule became increasingly untenable in the final quarter of the twentieth century. International ostracism, economic sanctions and external military involvement on the one hand and the growing internal political and economic pressures and problems on the other led to reappraisals in the 1980s. The morality of partition and secession was forcefully questioned (Buchanan 1991). Blatant racial domination similarly was condemned and viable alternatives were sought (Adam and Moodley

1993). However, official attempts aimed at the 'modernization' of apartheid under Prime Minister, later President, P.W. Botha, between 1978 and 1989 met with little acceptance. Relative White demographic decline, particularly highlighted by the massive rural–urban migration of the African population in the 1980s, illustrated the practical impossibility of implementing the entire apartheid policy. Furthermore, the escalating costs of maintaining the apartheid state resulted in severe economic problems, symbolized by the fall in the value of the national currency, particularly after the much publicized and defiant 'Rubicon' speech by President Botha in 1985 (Figure 0.3).[1]

The end was far more sudden than expected, following the epoch-making reversal of thought by President F.W. de Klerk in 1990. The massive legislative and administrative burden of the apartheid structure was dismantled in the course of 1991. The change in official thinking was summed up by the President in a condemnation of any consideration of retaining the system:

> Surely it is obviously unjust. Surely it is at variance with the Christian values we aspire to and profess. Surely this is in conflict with internationally acceptable norms. Surely this is a certain recipe for rebellion, civil war and revolution.
>
> (*Hansard* 1991: col. 7274)

The overwhelming rejection of apartheid by White voters in a referendum in 1992, and endorsement of a search for a negotiated settlement with the disenfranchised majority, prompted the President to state that 'South Africa has closed the book on apartheid' (South African Broadcasting Corporation television speech, 18 March 1992). After a series of tortuous negotiations the country's first universal-franchise elections in April 1994 led to the formation of the Government of National Unity under President Nelson Mandela. Politically, apartheid was formally abandoned.

However, the physical and social heritage of over forty-five years of enforced separation are overwhelming. The legacy of the policy has been a severe impediment to the country's development, as the Government of National Unity realized between 1994 and 1999. Indeed the Reverend Alan Hendrickse, then Leader of the Labour Party, suggested that

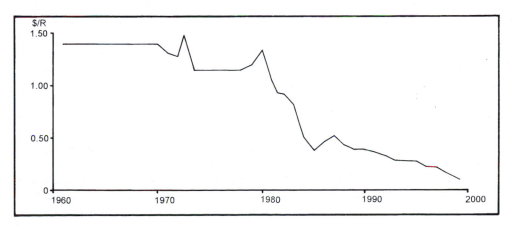

Figure 0.3 Value of Rand in US dollars, 1961–99
Source: Statistics from United Nations (1965–95) *Statistical Yearbook*, New York: United Nations, and *Sunday Times*, Johannesburg (1996–9)

even when pronounced dead, apartheid had the ability to remain alive, as the formal dis-mantling of the structure was not sufficient to undo the wrongs perpetrated in the era (*Hansard* 1992: cols. 5031–6). One perverse example has been the official continuation of racial classification in the interests of achieving affirmative action programmes and monitor-ing such legislation as the Employment Equity Act (1998), designed to reverse the effects of decades of discrimination in the workplace.

The period from 1948 to 1994 thus constitutes a remarkable period in world history, when South Africa occupied a particularly unenviable position as an international pariah state in consequence of its internal policies. *The Atlas of Changing South Africa* seeks to demonstrate the spatial patterns of the planning and enforcement of the policy and to examine the spatial heritage it has bequeathed as an indelible legacy for eradication in the new millennium.

State policy operated at three levels. 'Grand apartheid' sought to partition the country in an effort to ensure the continuation of White control in the remainder. 'Urban apartheid' sought to ensure that the living and business areas of the cities were segregated. Personal 'petty apartheid' sought to ensure minimal personal contact between people of different racial classifications, through aspects ranging from separate entrances to buildings to separ-ate schools. Each of these aspects was concerned with the control and use of physical space.

Politicians and bureaucrats drew lines upon maps at many scales in order to separate people into neat racial compartments. The essentially spatial characteristics of apartheid lend themselves to interpretation through the presentation of maps. Consequently in an era of rapid political change an atlas becomes a vital reference work for comprehending the nature of the transformation in progress.

The Atlas of Changing South Africa is not intended to be a chronological account of the history of South Africa since 1948. Many and conflicting accounts of this kind, have been and no doubt will be written on that subject (Thompson 1990; Saunders 1992, 1994; Saunders and Southey 1998; Smith 1988; Welsh 1998; Worden 1995). It is intended here to illustrate the spatial aspects of apartheid, demonstrating the essentially geographical plan-ning which was effected in the era. The work is therefore organized topically around broad themes in an attempt to bring together the various and diverse aspects of the South African state's apartheid project. Some of the more general tenets of apartheid will consequently be absent, where the inequalities are not spatially reflected, or where statistics are not available.

The collation of statistics on a racial basis, often excluding the African population or referring only to the White population, makes any work on South Africa particularly prob-lematical. Territorially, too, the term 'South Africa' has meant different extents, ranging from the broad concept of South Africa including Botswana, Lesotho and Swaziland, to the narrowly focused White apartheid state, excluding all the African homelands, whether 'independent' or not. In addition, for most of the period from the mid-1960s until 1994, the country was subject to international ostracism and economic sanctions. Thus much statistical material was regarded as sensitive and therefore not available to the researcher. Subsequent destruction of records has further reduced the possibilities of understanding the apartheid era.

If examples from Port Elizabeth appear frequently in the text, this is not because the city was the most segregated in the country or had any other distinctive feature, but because the author is most familiar with the peculiarities of the apartheid system as manifested in this city. Accordingly, it is hoped that readers may obtain a greater appreciation of the intricacies of this uniquely selfish and degrading policy and of modern South Africa in general through familiarity with a single city.

1 Before apartheid

The development of apartheid was firmly rooted in the colonial era. The country was first settled by European colonists under the Dutch East India Company in 1652. In 1806 the colony was occupied on a permanent basis by Great Britain. Although at the end of the Dutch period there were only 25,000 European settlers in the country, they had occupied land extending up to 1,000 kilometres from the original settlement at Cape Town through the evolution of an extensive farming system. Within the isolation which this system imposed, a distinct community evolved, later to become the Afrikaner nation. In the British period, South Africa was not perceived as a land of opportunity to rival the United States, Canada, Australia or New Zealand as a destination for emigrants (Christopher 1988a). As a result European immigration was never on a scale whereby British immigrants outnumbered the Afrikaners. Indeed between 1815 and 1914 less than 4 per cent of emigrants from the United Kingdom went to South Africa (Figure 1.1). However, the British contribution to the country in the administrative and commercial fields was profound.

Conflict between the British imperial authorities and sections of the Afrikaner community seeking political independence was a constant theme throughout the history of the nineteenth century. Thus on a political level the establishment of the Union of South Africa in 1910 as a British dominion was viewed as a compromise between the two White 'races', which consequently had nothing to offer the indigenous population in the form of political and economic prospects.

It would be a gross understatement to say that the indigenous population did not fare well under the colonial regime. The Khoisan peoples of the western half of the country, who came into contact with the colonists first, were worst affected. The population was sparse, as befitted a herding and hunting economy. Loss of lands and livestock as a result of the steady encroachment of the White colonists reduced most of its members to servitude, as servile tenant labourers on the newly established-European owned farms. Disease, notably smallpox, also took its toll, severely curbing population numbers in the first half of the eighteenth century. Thus in the western half of the country the indigenous population, reduced in numbers and deprived of land and livelihood, had shrunk to a small minority in many districts by the early twentieth century (Figure 1.2). In this respect the western half of the country followed the experience of many other mid-latitude colonies, notably the Americas and Australasia, where imported pathogens worked their deadly effects. However, in the eastern, better-watered, sector of the country the colonists encountered the Bantu-speaking peoples, with a mixed sedentary agricultural base. They were less susceptible to European diseases and proved to be difficult to control militarily, and so to dispossess of their land and livestock. Consequently, there ensued a long period of conflict from the 1770s to the 1870s,

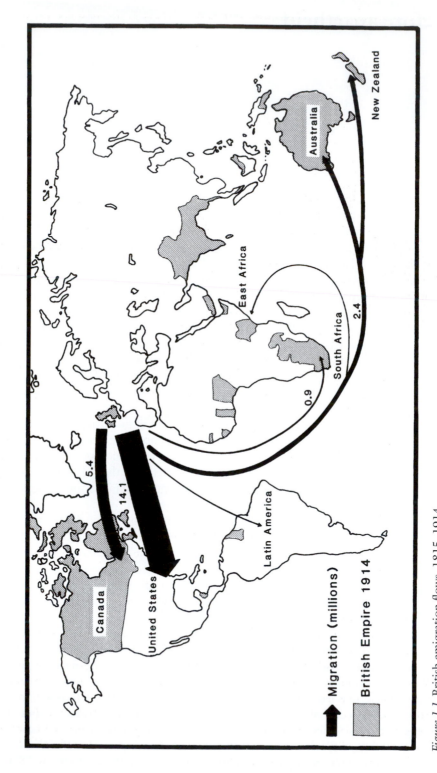

Migration (millions)

British Empire 1914

5.4

14.1

Canada

United States

Latin America

0.9

2.4

South Africa

East Africa

Australia

New Zealand

Figure 1.1 British emigration flows, 1815–1914
Source: From statistics in A.J. Christopher (1988) *The British Empire at its Zenith*, London: Croom Helm

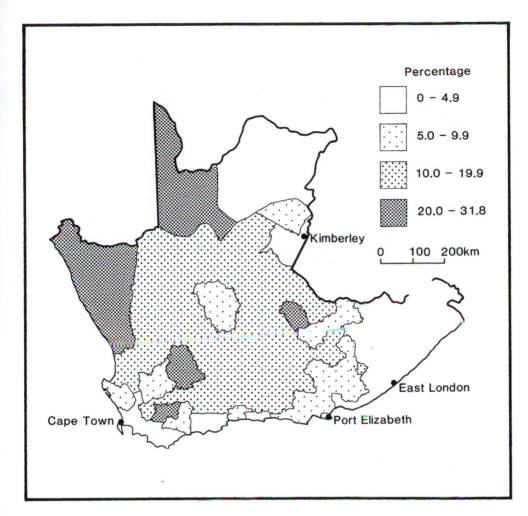

Figure 1.2 Distribution of Khoisan population, 1904
Source: Computed from Cape of Good Hope (1905) *Census of the Colony 1904*, Cape Town: Government Printer

while colonial and local governments conquered the various indigenous chieftaincies and kingdoms of the subcontinent in a piecemeal fashion (Mostert 1992).

In 1910 the Union of South Africa was established through the unification of four British colonies (Figure 1.3). At that time there were some five million people within the boundaries of the new dominion, of whom only one million were of accepted European origin. It was, however, this one million who controlled the government and remained determined to exercise exclusive political power in the ensuing eighty years. The methods employed for doing so changed from the pragmatic segregationism of the pre–1948 period to the ideologically formed apartheid policies of the post-1948 era. Although a distinct break at this date is often implied, there is a remarkable degree of continuity from the colonial period through the Union segregationist era to the age of apartheid.

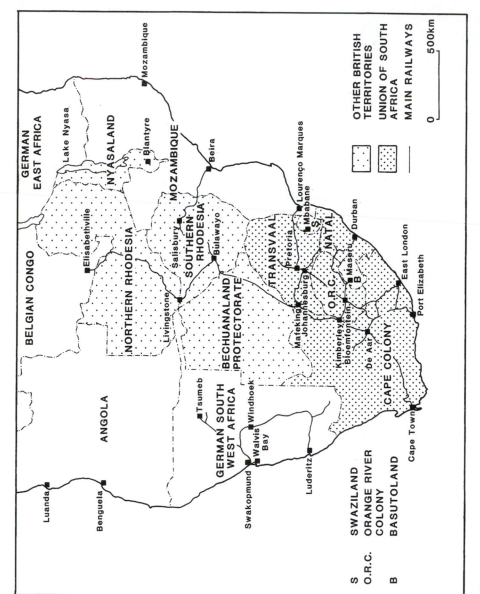

Figure 1.3 Southern Africa, 1910
Source: Compiled by the author from various sources

COLONIALISM

The maritime colonial powers initially bypassed South Africa in their quest for power and profit in the East. The Portuguese, though mapping the coastline, did nothing to establish permanent bases on this section of the continent, as there were no recognizable exploitable resources and the indigenous population appeared hostile. Only when the advantages of establishing a provisioning station for ships voyaging between Europe and the East Indies were perceived did the Dutch East India Company decide to act. In 1652 a settlement was founded at Cape Town to provide fresh fruit, vegetables and meat to the Company's vessels (Christopher 1976). An extensive garden was laid out to serve these needs. However, the Company's limited resources were such that it was not possible to fulfil the expectations, and recourse was made to the settlement of free individual European settlers to grow the required provisions. The supply of meat was also problematical as the indigenous population proved reluctant to barter their cattle away in the numbers required. Conflict and seizure ensued, with the Company, and later individual settlers, undertaking cattle, sheep and goat raising on lands taken from the indigenous population. Attempts to define the physical boundary of the settlement, in order to limit the extent of the conflict, began with the planting of a hedge of bitter almonds instigated by Jan van Riebeeck, the first Dutch commander (Figure 1.4). This, like all subsequent demarcations, proved to be ephemeral as European settlers seized the opportunities which were offered by a new land with minimal government interference and an initially limited indigenous population.

The colonial scale of operations thus increased rapidly as extensive systems of cultivation and stock farming replaced inherited intensive European models. A constant lack of labour and capital hampered development, as all the colony appeared to offer in abundance was land. Settlers from the Netherlands, Germany and France were placed on small (20–50 ha) farms in the seventeenth century. However, in 1717 the government halted free immigration as the colony was considered to be fully occupied, although it contained only 2,000 European inhabitants. The limited physical resources of the immediate hinterland of Cape Town restricted commercial agriculture to grain and wine enterprises, which were run on tropical plantation lines with slave labour forces imported from the Dutch colonies in Asia. The vital decision as to whether to run the colony with free or servile labour was taken in 1658, effectively determining the form of colonial society, and thereby influencing the social attitudes and population composition of later generations.

Parallel with these developments was the evolution of extensive livestock raising in the interior of the colony. The seasonal movement of livestock to lands in and beyond the mountains of the western Cape, evolved by the early eighteenth century, was subsequently transformed into the permanent establishment of independent stock farms in these regions. Farms of 2,500 hectares were offered at a nominal rental to those venturing on to the frontier of European settlement. The result was the rapid appropriation of the best agricultural and grazing land across the subcontinent and the dispossession of its occupants (Figure 1.5). By the late eighteenth century lands were being settled over 1,000 kilometres from Cape Town, and the settlers had come into conflict with the indigenous agriculturalists, whom they were unable to displace on any significant scale with the technology available to them. Settler pastoralism differed little from that of the indigenous population whom they incorporated into the system wherever possible. Opportunities for commercialization were limited as distance precluded the marketing of more than small quantities of livestock and livestock products in exchange for luxury items, firearms and wagons. In the main, settler society was self-contained with only limited contacts with the

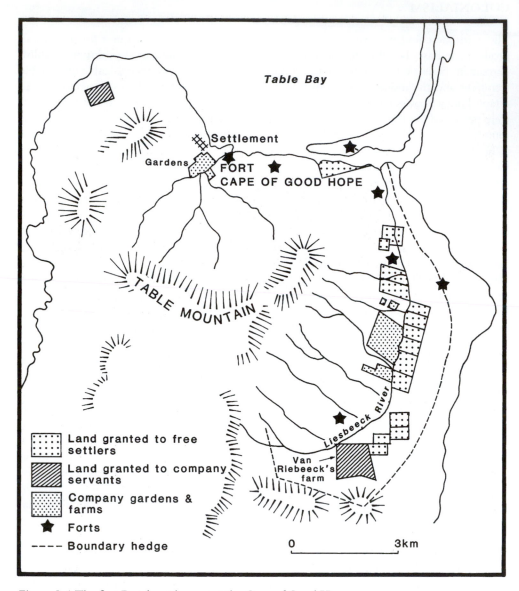

Figure 1.4 The first Dutch settlement at the Cape of Good Hope
Source: After A.J. Christopher (1976) *Southern Africa*, Folkestone: Dawson

commercial world or indeed between individual members. Isolation was the overwhelming result of the extended lines of communication to and from Cape Town and the poverty of the physical environment (Lamar and Thompson 1981).

In the nineteenth century the process of frontier extension was continued. Afrikaner settlers pushed into the interior of South Africa across the Orange River, and entered areas initially considered to be virtually devoid of indigenous people as a result of the massive upheavals and migrations of the era of the Difaquane, or 'forced migration', associated with the rise of the Zulu military monarchy in the 1820s and 1830s (Wilson and Thompson

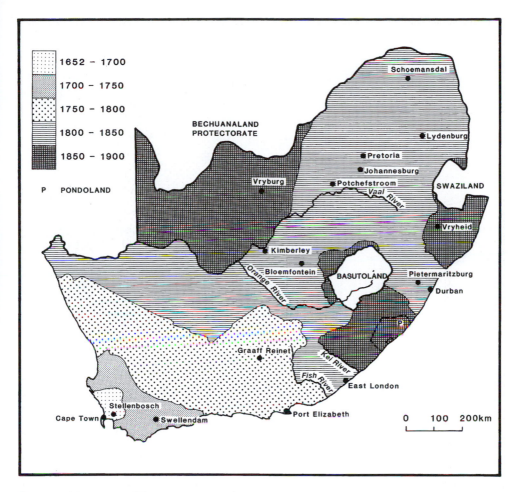

Figure 1.5 Sequence of European annexation
Source: Compiled by the author from various sources

1969: 391). In the late 1830s some 15,000 Afrikaner colonists migrated north and east-wards to occupy large tracts of the Orange Free State, Transvaal and Natal. In these regions they established their own independent republics, free from British control. Chaos in Natal led to British annexation and settlement by new immigrants, but the Orange Free State and Transvaal (South African Republic), after involved political manoeuvring, survived as separate entities until the end of the century, when they again came under British rule. The Anglo-Boer War (1899–1902) is one of the epic conflicts in the building of an Afrikaner mythology and sense of identity (Pakenham 1979). The political map of 1899 is thus still able to evoke political emotions as representing the Afrikaner heartland (Figure 1.6).

Within the republics land was appropriated by the settlers in a similar fashion to that evolved in the Cape Colony. The same semi-self-sufficient economy was reproduced, although ivory possibly contributed more to local funds than agriculture as such in the first decades of settlement. Commercialization of farming was only possible with the develop-ment of wool exports in the 1830s and 1840s in the Cape Colony, and the extension of

Figure 1.6 Political map of South Africa, 1899
Source: Compiled by the author from various sources

sheep farming elsewhere in the ensuing decades. However, in much of the Transvaal sheep were subject to diseases which precluded their commercial exploitation and no viable substitute was introduced until the twentieth century.

Within these new states and colonies the indigenous population remained in substantial numbers, both on the periphery, where effective settler control was only established late in the century, and in the European-occupied farming area, where substantial workforces were acquired along with the land. So large were the indigenous populations incorporated within colonial and republican state boundaries from the 1830s onwards that a new form of organizational control was introduced. Rural reserves were set aside where the indigenous population could continue to live under traditional rulers and customary laws but subject to overall European supervision. In some cases the lands were granted to co-operative African leaders who assisted the settlers or the governments, while in other cases they were demarcated to accommodate those considered to be 'surplus' to the labour requirements of the settlers.

Later, as the colonies and republics incorporated regions which were densely occupied by the indigenous population, little change was made in the land dispensation as virtual protectorate status was accorded to the indigenous social hierarchy. In all cases it needs to be emphasized that the land so set aside was regarded as belonging to the government, to be taken away if the conditions of agreement were infringed. No indigenous land rights were recognized as the basis of a colonial or republican title deed. Conquest and the act of legal annexation effectively transferred sufficient rights to the Crown or the state to ensure that all further land holdings derived their rights from those authorities, the colonial *terra nullius* principle. However, attempts to draw a permanent line to define the boundaries between immigrant and indigene remained a constant theme of nineteenth-century politics, as they had done since the foundation of the first Dutch colony, and were to become an integral part of the apartheid policy.

THE POPULATION

The population of South Africa is highly complex both in its origin and in its distribution. The indigenous population, broadly divided into the Khoisan population of the western regions, and the Bantu-speaking population of the eastern regions, constitutes the numerical majority, while the immigrant peoples are most highly concentrated into areas of specific economic opportunity, notably the towns and restricted rural areas outside the defined reserves (Figure 1.7).

In any discussion of population in South Africa the racial connotation is of paramount importance, as membership of a particular racial or ethnic group has largely defined the individual's opportunities and rights (Elphick and Giliomee 1979). Thus throughout the colonial and Union periods racial classification was of the utmost significance. At first the basic distinction was made between Christian and Heathen. Later this was amended to Free and Slave, then the terms 'European' and 'Coloured' were adopted to denote the same divide. Such bi-polar divisions became increasingly fragmented in the nineteenth century when, under the influence of anthropological taxonomy, ever more elaborate classifications were used in the censuses. Accordingly, in the last Cape Colonial census in 1904 seven major categories were employed, fragmented into over forty subdivisions. Although simplified at Union, the census commissioners continued to employ a racial classification system, based on physical and social characteristics. The problems of such a classification may be illustrated by the grouping 'other, mixed and other' in 1904 (Cape of Good Hope 1905). Broadly some four groups were distinguished in the Union period, namely: White or European, Native (later Bantu, Black or African), Coloured or Mixed, and Asian. As evidenced in other classificatory censuses, notably in British India, the categories devised for taxonomic purposes assumed legal and political status (Barrier 1981).

The White population of European origin constituted a minority throughout South Africa by 1951 (Figure 1.8). It was derived from a range of countries, but broadly fell into two groups, the Afrikaans and English speakers, who controlled the political process. Substantial German and Portuguese communities were also present. It should be remembered that the majority (approximately 55–60 per cent) of the European population was Afrikaans speaking, and that between 1910 and 1994 all South African prime ministers and presidents were drawn from that community (Figure 1.9). In contrast the cities, because of the pattern of colonial economic development, were largely dominated by English speakers until the

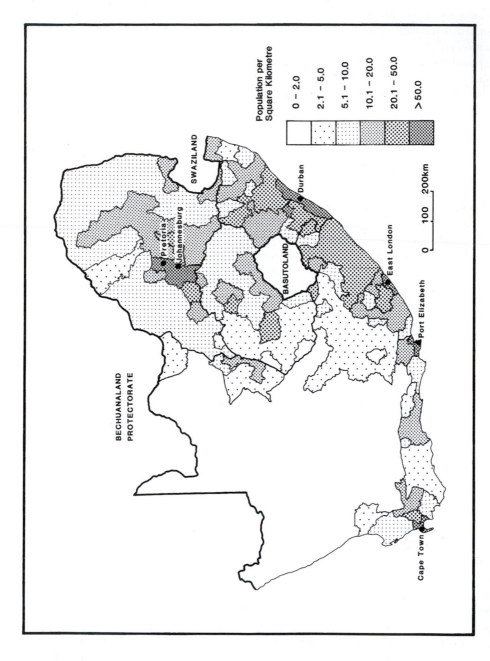

Figure 1.7 Distribution of population, 1951
Source: Computed from South Africa (1955) *Population Census 1951*, Pretoria: Government Printer

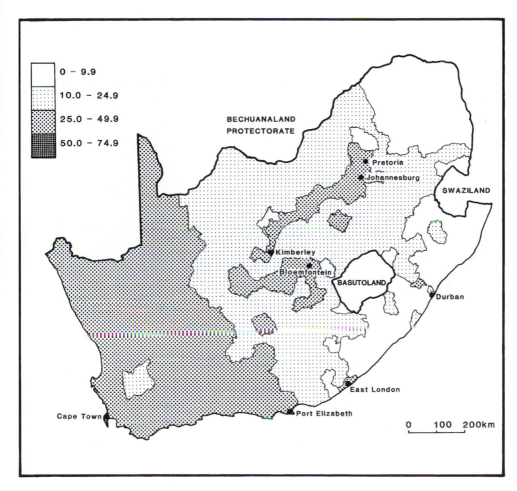

	0 – 9.9
	10.0 – 24.9
	25.0 – 49.9
	50.0 – 74.9

BECHUANALAND
PROTECTORATE

Pretoria
Johannesburg

SWAZILAND

Kimberley
Bloemfontein

BASUTOLAND

Durban

East London

Cape Town
Port Elizabeth

0   100   200km

Figure 1.8 Percentage of population returned as White, 1951
Source: Computed from South Africa (1955) *Population Census 1951*, Pretoria: Government
Printer

1950s. The point must be emphasized that this segment of the population no longer viewed itself in colonial, settler, terms, but as permanent 'South Africans' (Adam 1994).

The numerically dominant, officially defined indigenous Native, African, Bantu or Black population was composed only of Bantu-speaking people, who had not settled in the Western Cape by the beginning of the colonial era (Figure 1.10). Owing to the long history of occupation the various subgroups developed separate languages, often mutually unintelligible (Inskeep 1978). Three broad ethno-linguistic divisions are represented in the country: the Nguni peoples of the eastern coastal regions, the Sotho peoples of the interior, and the Vendans and Shangaans of the extreme north and east of the Northern Province. In precolonial times these peoples were subject to a range of authorities from the centralized Zulu monarchy to the individual village headmen, subject to only a tenuous higher authority.

The Asian population was largely descended from Indians who initially came as indentured labourers to work on the sugar plantations of Natal in the nineteenth century. Later,

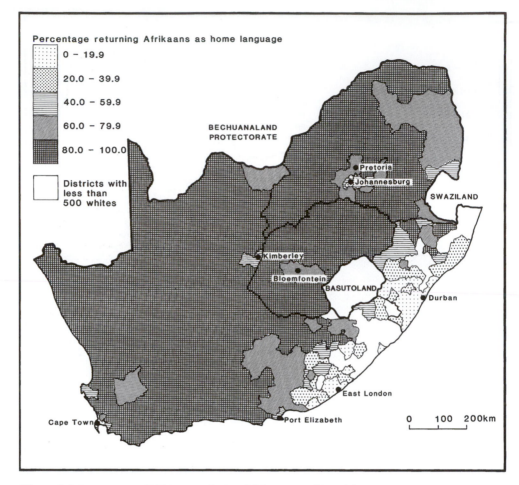

Figure 1.9 Percentage of White population Afrikaans speaking, 1946
Source: Computed from South Africa (1954) *Population Census 1946*, Pretoria: Government Printer

others immigrated from that subcontinent as traders and professionals to serve the community, which consequently reflects the religious, linguistic and cultural diversity of India. As in other British self-governing colonies the remarkable commercial enterprise of some members of the Indian community prompted hostility from European commercial interests which were perceived to be threatened by industrious competition (Huttenback 1976). The experiences of Mahatma Gandhi in South Africa helped to mould his philosophical approach to the movement which ultimately led to the independence of India (Swan 1985). The community was predominantly urban by 1951 and so constituted a significant part of the population of the major centres of Natal and the southern Transvaal and to a lesser extent the Cape coastal cities (Figure 1.11). Smaller numbers of Chinese emigrated to South Africa between 1890 and 1949 and form a distinct subgroup largely engaged in trade (Yap and Man 1996).

The Coloured population was a general category devised to encompass everyone not

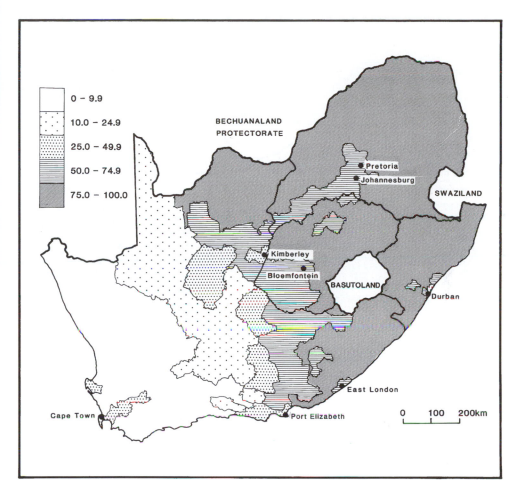

Figure 1.10 Percentage of population returned as African, 1951
Source: Computed from South Africa (1955) *Population Census 1951*, Pretoria: Government Printer

catered for in the above categories. Spatially it was concentrated in the Western and North-ern Cape (Figure 1.12). The main component was made up of the people of mixed Euro-pean ancestry, who were socially excluded from being accepted as European. Also included were the descendants of the slaves imported into the Cape Colony by the Dutch in the seventeenth and eighteenth centuries. They, the Cape Malays, were acculturated to the extent of adopting Afrikaans in speech, but spiritually found their solace in Islam (da Costa and Davids 1994) (Figure 1.13). On the basis of acculturation the descendants of the Khoisan population were also included, as they were also incorporated linguistically into the Afrikaans-speaking population. The Coloured community thus included a wide diversity of origins, which varied according to the region concerned.

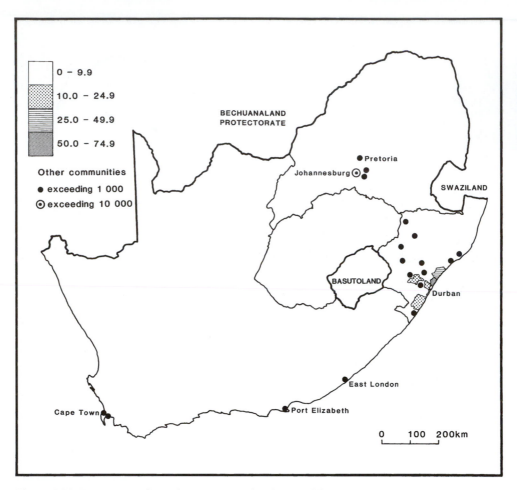

Figure 1.11 Percentage of population returned as Asian, 1951
Source: Computed from South Africa (1955) *Population Census 1951*, Pretoria: Government Printer

THE ECONOMY

The colonial economy was dominated by the international market and its links to the interior of the subcontinent. Until the nineteenth century the value of South African imports and exports was extremely limited and the country could be described as essentially peripheral (Nitz 1993). Agricultural products, notably wine, wool and ostrich feathers, dominated the export trade until the 1870s (Figure 1.4). Thereafter minerals, first diamonds and then gold, assumed the dominance which they maintained for over a century (Browett 1976). This is not to suggest that colonial society was completely commercially oriented and linked to the international economy. It has been noted that parts of the White agricultural sector were only commercialized during the Second World War.

The agricultural sector played a vital role in the development of the country before 1948. The wealth generated by exports of wine and grain had given impetus to the emergence of a

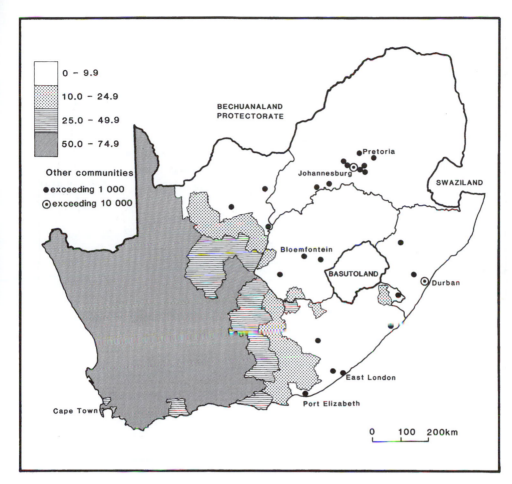

Figure 1.12 Percentage of population returned as Coloured, 1951
Source: Computed from South Africa (1955) *Population Census 1951*, Pretoria: Government Printer

rural gentry by the mid-eighteenth century (Guelke and Shell 1983). In the nineteenth century the development of pastoral woolled sheep, angora goat and ostrich farming introduced commercialization to a wider extent of the country. In the twentieth century maize and sugar cultivation diversified exports and led to commercialization of more regions. Internal cattle and grain markets provided a further underpinning of the economy. However, by 1910 agriculture only contributed 17 per cent of the national income, a figure reduced to 13 per cent forty years later (Figure 1.15).

The development of the mining industry in the last third of the nineteenth century transformed the national economy dramatically. Revenues generated by the mining of diamonds at Kimberley and gold on the Witwatersrand were invested in massive infrastructural and conspicuous spending. More particularly, industrialization which had begun in the coastal port cities in the 1850s proceeded rapidly. Urbanization on a substantial scale was initiated with a demand for labour in the mines and the attendant services. Although initially

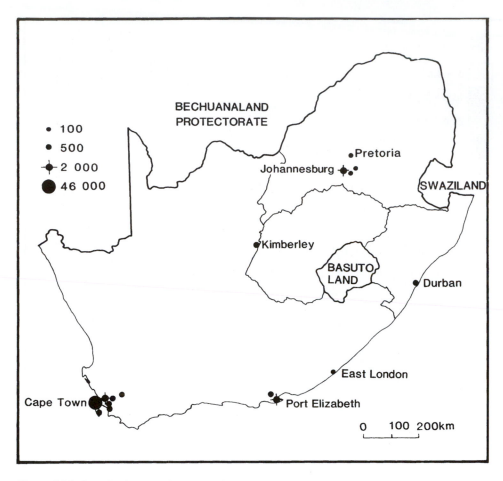

Figure 1.13 Cape Malay population, 1951
Source: Computed from South Africa (1955) *Population Census 1951*, Pretoria: Government Printer

much of the labour in mining was White, soon the heavy tasks were undertaken by Africans. The growing demand for more labour, particularly on the gold mines, was such that mine owners had to resort to large-scale recruitment from further and further afield. In the 1890s labourers were being recruited from neighbouring countries, including Mozambique. By the inter-war period all the African countries south of the Equator were represented in the South African mines.

The mining industry introduced large-scale migrant labour systems. African miners were not recruited as permanent workers and therefore did not bring their families with them to live on the mining settlements. Instead the compound system, evolved for the specific security purposes on the diamond mines at Kimberley, was adopted as the norm. Accordingly, large hostels for African adult males were constructed adjacent to the mine but separate from the remainder of the mining settlement provided for the permanent White (and Coloured) staff. The mining companies therefore did not have to bear the costs of supporting the families of the migrants, which were borne by the agricultural sector in the rural reserves.

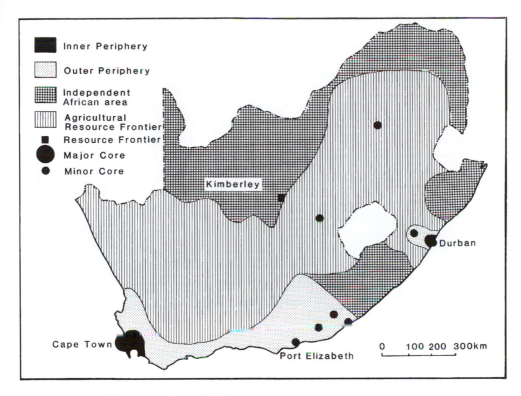

Figure 1.14 Development regions, 1870
Source: After J.G. Browett (1976) 'The application of a spatial model to South Africa's development regions', *South African Geographical Journal* 58: 118–29

Mining also transformed the space economy of the country. Both major mineral discoveries were made on what were then peripheral parts of the settler economy. Kimberley was situated on the then frontier of European settlement, but the diamond deposits were comparatively limited in extent. The Witwatersrand gold discoveries led to the establishment of a series of new towns extending 80 kilometres from Krugersdorp in the west to Springs in the east, which became the hub of the national economy. The main settlement, Johannesburg, though only founded in 1887, was by 1911 the largest city in the country (van Onselen 1982). The isolated coastal economies were then linked up to the Witwatersrand, which became the focus of the rail and road systems.

Attendant upon the development of mining was the expansion of basic industry and the manufacturing sector. At first agricultural processing had formed the basis of much local industry. To this was added basic metal work for the mining industry. This in turn led to the demand for the establishment of an iron and steel industry, which was only successfully undertaken by the state through the establishment of the South African Iron and Steel Corporation in 1928. State direction resulted in the selection of the capital, Pretoria, as the site of the first plant and the careful integration of further plants within national planning policies. Manufacturing industry received a major boost in the Second World War when efforts to achieve a degree of independence from uncertain overseas supplies resulted in

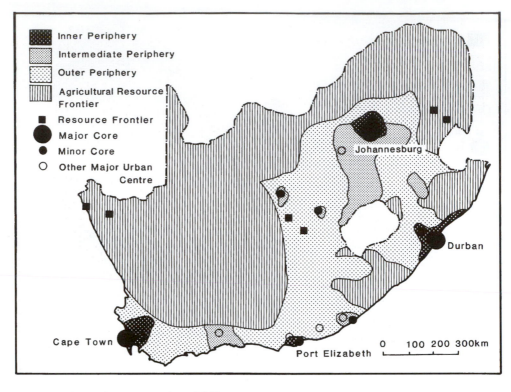

Figure 1.15 Development regions, 1911
Source: After J.G. Browett (1976) 'The application of a spatial model to South Africa's development regions', *South African Geographical Journal* 58: 118–29

the sector overtaking mining as the most significant contributor to the national income (Figure 1.16).

The boom in the economy associated with the Second World War resulted in a further impetus to urbanization. African rural–urban migration was on such a scale that by 1946 there were more Africans living in the urban areas than Whites. The dominance of the metropolitan area extending from Pretoria to the Vaal River was well established by 1960 and accounted for nearly half the country's net output and employment in manufacturing (Figure 1.17). Such a concentration of economic opportunity attracted rural as well as urban migrants. The systems of control to prevent African urbanization proved to be ineffective, while refusing to recognize the enormity of the problems raised by inadequate housing and services only resulted in confusion in official circles and extensive slum conditions. In 1948 a United Party government commission of inquiry recognized that the African community was a permanent part of the urban population of South Africa (South Africa 1948). It further noted that the national economy was racially integrated and that the African community formed an integral part of it. This report was in marked contrast to the prevailing concept that the towns were essentially White, both in control and occupation. Indeed a strong unifying political thread from the nineteenth century, in all parts of the country, had been the provision of separate urban locations or townships for the indigenous population, where social separation was reinforced by physical separation.

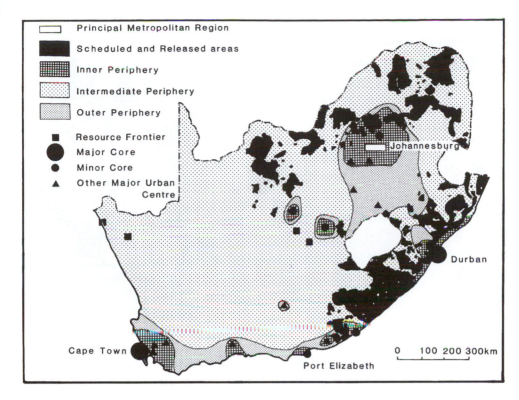

Legend:
- Principal Metropolitan Region
- Scheduled and Released areas
- Inner Periphery
- Intermediate Periphery
- Outer Periphery
- ■ Resource Frontier
- ● Major Core
- • Minor Core
- ▲ Other Major Urban Centre

Figure 1.16 Development regions, 1936
Source: After J.G. Browett (1976) 'The application of a spatial model to South Africa's development regions', *South African Geographical Journal* 58: 118–29

THE POLITICAL DISPENSATION

The political development of South Africa prior to 1948 was complex. The Cape Colony was administered by the Dutch and later the British until 1910, although self-government was introduced in 1853. The republics established in the Transvaal and Orange Free State by Afrikaner trekkers, escaping from British rule in the 1830s, were recognized as autonomous states by Great Britain in the 1850s. Natal, however, was annexed to the Crown in 1843. Later, ephemeral republics and colonies on the frontiers of European settlement were established, but by 1900 they had been incorporated into the four adjacent states. The anomalous survival of Lesotho (Basutoland) and Swaziland as separate entities dates from this period of flux in colonial administrations, which became fossilized in 1910. Accordingly, in 1910 the Union of South Africa was formed from the four self-governing colonies of the Cape, Natal, Transvaal and Orange Free State.

Control of the new British dominion was vested in a government elected by parliament. Parliament itself was controlled by the White population, through the exclusion of virtually all the remainder of the population from the franchise. Each province retained the form of franchise it had enjoyed in the late colonial period. Thus in the Orange Free State and the Transvaal universal White adult male franchise endured. In the Cape Province and Natal a

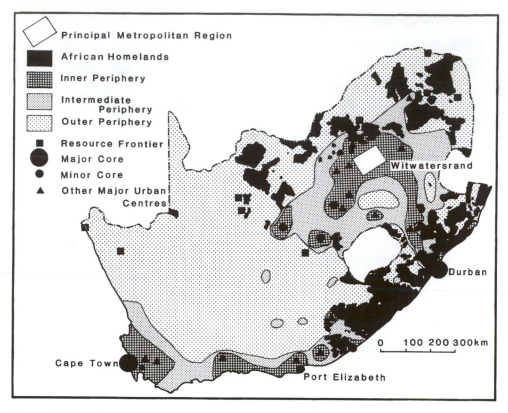

Figure 1.17 Development regions, 1960
Source: After J.G. Browett (1976) 'The application of a spatial model to South Africa's development regions', *South African Geographical Journal* 58: 118–29

qualified franchise open to all was adopted, subject to property and income qualifications. In Natal the franchise had been restricted in such a way as to exclude the indigenous population and virtually the entire Indian population. In the Cape Province there was a much more open approach with a substantial Coloured and African vote. Regarding the distribution of seats between the provinces, however, only White voters were counted constitutionally, leading to a built-in under-representation of the Cape Province throughout the Union period. In 1932 when women gained the vote, it was restricted to Whites only, thereby halving the impact of the Cape Coloured vote. In 1936 Africans were removed from the voters' roll and Senators appointed to oversee African interests. Thus the South African state was firmly administered according to laws which served the interests of the White population.

   Politically, through malapportionment and the gerrymandering of the colonial nineteenth-century republican and Union electoral systems, the farming community dominated local parliaments until after the National Party victory in 1948. Thus policies demanded by the farming community were sympathetically addressed by successive governments. Furthermore, the gerrymandering in favour of the rural areas ensured a disproportionate Afrikaner influence on parliament (Figure 1.18). Through much of the

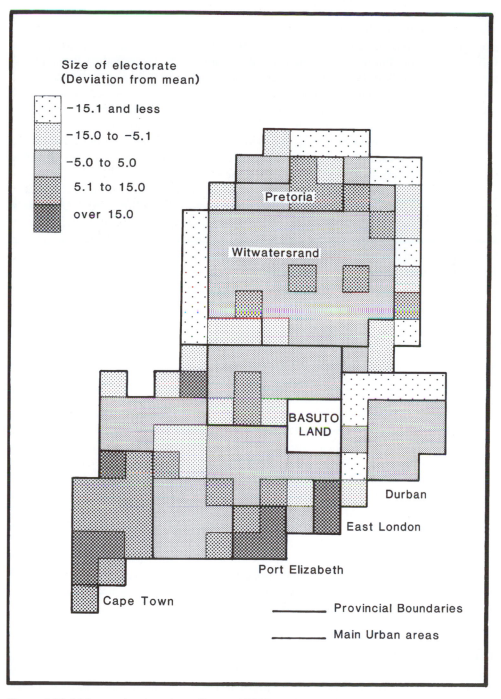

Figure 1.18 Malapportionment in parliament, 1948
Source: Drawn from statistics in South Africa, Report of the Ninth Delimitation Commission, *Government Gazette* 3931, 13 February 1948

Union period the English-speaking voter had little influence on the course of government policy.

Between 1910 and 1948 there were only three prime ministers, all of whom had been Boer generals in the Anglo-Boer War of 1899–1902. They formed several administrations of various persuasions but with a consistent segregationist policy. The founding Prime Minister, Louis Botha (1910–19), and his successor, Jan Smuts (1919–23 and 1939–48), stood for reconciliation between the Afrikaans- and English-speaking communities, while J.M.B. Hertzog (1923–39) was more concerned with developing a South African nationalism based on the concept, 'South Africa first'. All three were interventionist, placing considerable volumes of legislation on the statute book to regulate the relationship between the races, which each regarded as being one of the major areas of government concern, as well as numerous laws having an indirect bearing upon the subject.

White political parties tended to revolve around personalities and splits in the Afrikaner vote (Davenport 1991). The broad grouping which General Louis Botha led at Union split with General Hertzog's establishment of the National Party in 1913. In 1923 General Hertzog was able to form a government with the support of the Labour Party, which was particularly strong among English-speaking White mineworkers on the Witwatersrand. Racial protection through urban segregation and industrial job reservation for Whites provided a joint programme for the two parties. In 1933 the world economic crisis led to a political accommodation between Generals Hertzog and Smuts to form the United Party. A splinter group of Nationalists under Dr D.F. Malan established a rival Purified (later Reunited) National Party and proceeded to gain Afrikaner support until gaining power in the election in 1948. In 1939 General Hertzog resigned over the issue of entering the Second World War in support of Great Britain, which he opposed. General Smuts returned as prime minister to pursue the war and play a prominent role in international affairs.

Control of parliament was particularly important in South Africa as the economy was characterized by a high degree of state intervention. State ownership of the infrastructure, including the railways and harbours, resulted in an integrated transport policy. State control of key industries determined their location and hence their role in the urbanization process. Through active fiscal and rating policies, certain branches of local industry were fostered and agriculture protected. The African labour force on both the farms and the mines was regulated and controlled by a complex series of laws restricting individual movement. Town planning legislation was used as an instrument of control to limit urban African numbers and confine them to specific parts of the town, just as rural policies confined much of the African population to specific reserves. The South African colonial and post-colonial economies and societies were thus highly regulated by the government. The apartheid era was accordingly a continuation of the state interventionist policies of the previous era. It was the scale of intervention and the ruthlessness with which it was undertaken which distinguished the post-1948 era from that which preceded it.

THE LAND ACTS

The division of rural land between the African and White communities had largely been made by the time of Union in 1910. One of the first actions of the Union government was to effect a legal division describing those areas of the country which were assigned to the indigenous population. Under the Natives Land Act of 1913 some 8.9 million hectares were defined as Native reserves (Figure 1.19). Furthermore, the Act prohibited the purchase of

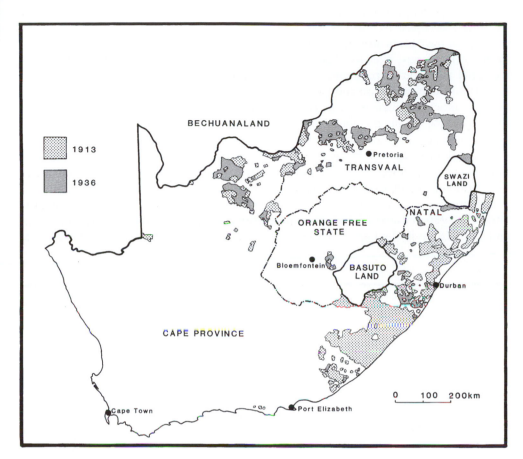

Figure 1.19 Areas designated under the 1913 and 1936 Land Acts
Source: Redrawn from official 1:250,000 topo-cadastral maps, various dates, Pretoria: Government Printer

land by members of the African community outside the scheduled reserves, except in the Cape Province. Provision was also made for the prevention of other means of independent African access to land, through squatting, leasing, share-cropping or labour tenancy. The latter provision was resisted in many parts of the country by White landowners as they envisaged the system as the only means of overcoming a perceived dearth of labour. However, measures directed against share-cropping and labour tenancy agreements were locally enforced, resulting in the initiation of the major resettlements associated with the removal of all Africans not directly dependent upon White employers (van Onselen 1996). The map depicts only the formal position, not the multitude of informal arrangements which were effectively invalidated by the Act (Horn 1998). Nor does it reflect the complex structures within the designated areas (Murray 1992).

The Natives Land Act was officially conceived as a first stage in drawing a permanent line between Africans and non-Africans. A subsequent commission of inquiry (Beaumont Commission) reported in 1916 that extensive additional lands should be added to the reserves, which were already exhibiting signs of impoverishment (Bundy 1979). Noting that

3.6 million hectares of White-owned land and 0.8 million hectares of Crown land were exclusively occupied by Africans, the commission suggested that they should be legally included in the African areas, and additional provision made for those being displaced from White-owned farms. The First World War and the subsequent political turmoil delayed any legislative action on the recommendations for twenty years.

In 1936 the Native Trust and Land Act finally provided for the extension of the African areas by some 6.2 million hectares. Here it should be noted that in most cases the schedule attached to the Act did no more than recognize the existing pattern of African occupation, already noted by the Beaumont Commission. In terms of spatial patterns the broad consolidating additions suggested by the Beaumont Commission were rejected in favour of a more fragmented distribution to include individual farms occupied by Africans (Figure 1.20). Furthermore, the implementation of the Act involved the purchase of the land through parliamentary grants, which was to be a slow process, only completed more than forty years later. The 1936 Act also took away the right of Africans to purchase land outside the reserves and Trust Lands in the Cape Province, and in a parallel act, African voters were removed

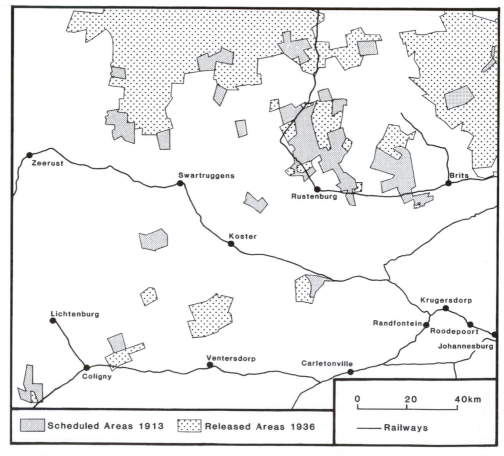

Figure 1.20 Western Transvaal: areas designated under the 1913 and 1936 Land Acts
Source: Redrawn from official 1:250,000 topo-cadastral maps, various dates, Pretoria: Government Printer

from the multi-racial Cape voters' roll (Tatz 1962). A South African Native Trust, subsequently renamed several times, was created to be the controlling agent of land purchased
under the Act. The legislation was constantly amended to enforce the rules adopted to
achieve the removal of Africans from the White rural areas, except those working as paid
employees, as the concept of segregationism became ever stronger in governmental
thinking.

It should be noted that the Native reserves and the Trust Lands accommodated less than
half the total African population during the Union period. In 1913 approximately half the
African rural population lived on the proclaimed reserves and a further sixth on Mission
stations, African-owned farms, and lands exclusively occupied by Africans. The proportion
fell slowly in the Union period. In 1960 just under half the African rural population lived on
all categories of designated African areas. However, as a result of the migration to the towns,
only a third of the entire African population continued to live in the reserves and Trust
Lands.

URBAN SEGREGATION

Formal urban residential segregation began in the aftermath of the abolition of slavery in
1834 and the first substantial influx of Africans to the towns. Accordingly, in 1834 the
London Missionary Society in Port Elizabeth established a separate suburb for its charges,
physically isolated from the remainder of the town (Christopher 1987). It was to be first of
several peripheral suburbs, setting a precedent for the 'locations' which were to be established in South Africa in the ensuing 150 years (Figure 1.21). In 1847 the Cape colonial
government promulgated regulations for the establishment of separate locations close to
existing (White) towns where members of indigenous groups, including Afrikaans-speaking
Coloured people, were required to live, if not independent property owners themselves or
housed by their employers. The result was the establishment of subsidiary residential settlements throughout the eastern Cape in the second half of the nineteenth century. The
process received added impetus as a result of the major bubonic plague outbreak of 1901,
when the White demand for segregation was reinforced by perceived health hazards, creating the 'sanitation syndrome' and a new generation of even more peripheral African settlements (Swanson 1977). It should be noted that in the Cape Colony and Natal no racial
restrictions were legally imposed on African and Coloured ownership or occupation of land
until the late nineteenth century. However, few members of these two communities possessed the financial means to avail themselves of the legal opportunities offered to them.

The Orange Free State and Transvaal governments adopted the Eastern Cape system of
separate residential areas for the African and Coloured populations. The degree to which
enforcement took place varied from the strict regimes of the Orange Free State to relatively
ineffective enforcement in the majority of Transvaal towns before 1900. The Natal authorities effected influx control measures to prevent a permanent African presence in the towns,
relying upon a migratory system to meet urban demands for labour.

The mining companies introduced segregated compounds to house African single
migrant workers in the settlements under their control (Mabin 1986). Strict supervision of
such compounds was introduced to overcome the problem of illicit diamond smuggling in
the Kimberley area (Figure 1.22). However, the degree of control which could thereby be
exercised over the costs of labour and over the miners' activities made the system attractive
for the subsequent gold and coal mine owners who did not have the same security problem.

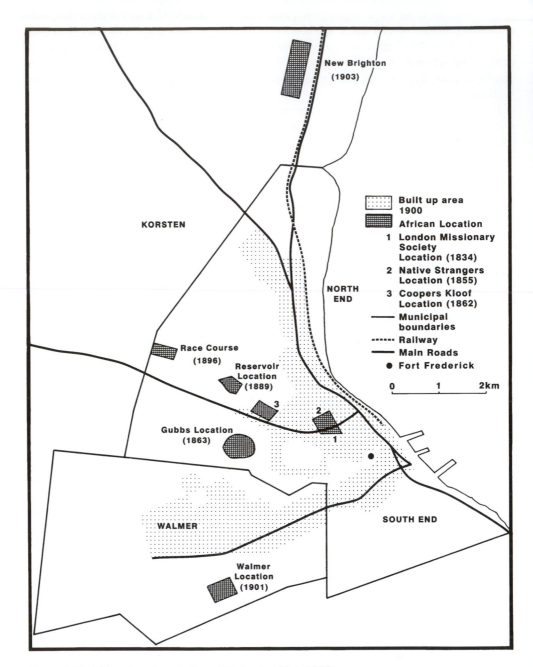

Figure 1.21 African locations in Port Elizabeth, 1834–1903
Source: After A.J. Christopher (1987) 'Race and residence in colonial Port Elizabeth', *South African Geographical Journal* 69: 3–20, and Map of Walmer (1926) Walmer Municipality

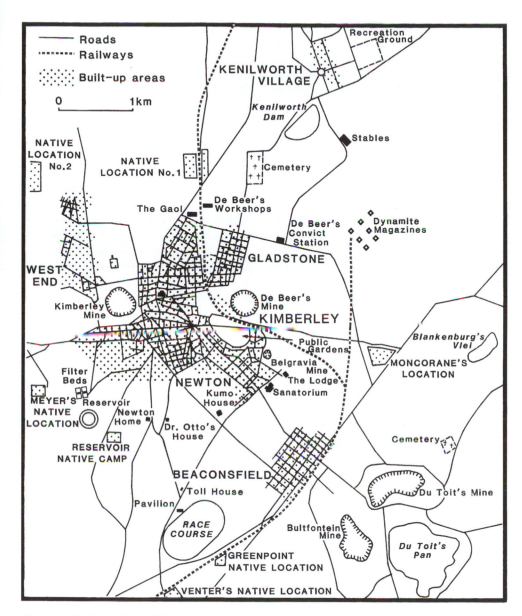

Figure 1.22 Kimberley, 1900
Source: After A.J. Christopher (1976) *Southern Africa*, Folkestone: Dawson

Thus the steady decline in mining wages in the late nineteenth and early twentieth centuries has been attributed, in part, to the settlement form adopted to control the labour force.

At the time of Union a wide variety of urban policies relating to the indigenous population were in force, in contrast to the reasonably uniform rural reserve policies (Christopher 1988b). The attempt to introduce national uniformity was only undertaken in the 1920s, as a result of the accelerated migration of Africans into the towns and the consequent competition for jobs between African and White workers (Davenport 1971). In 1922 the Transvaal

Local Government Commission recommended a general tightening of regulations and the entrenchment of the concept of the urban areas as 'the domain of the White Man' in which 'there was no place for the redundant Native' (Transvaal 1922: 42).

In 1923 the Natives (Urban Areas) Act was passed requiring local authorities to establish separate locations for the African population, and to exercise a measure of control over the migration of the African population to the towns (Davenport 1970). The Act was amended and tightened regularly, notably in a major consolidating act in 1945, which formed the basis of the apartheid era legislation.

A number of municipalities resisted the development of locations, questioning the expense involved. Durban Municipality, for example, did not build its first location until the 1930s, preferring that employers, whether private individuals, companies or governmental organizations, undertake the financial responsibility (Torr 1987). Municipal councils and their predominantly White electorates were reluctant to spend money from the general rate fund on the construction and administration of African locations. The 'Durban system' of funding the construction of houses and service infrastructure through the sale of indigenous beer was widely adopted, but severely limited the sums available for development (Swanson 1976). Hence, in contrast to the White suburbs, the African areas were remarkably devoid of services, a legacy still in evidence.

By 1948 the system of physically separate African locations was firmly in place throughout the country. Indeed the majority of towns had established African locations by the time of Union, although those in the western Cape essentially housed the Coloured population at that date and therefore possessed no formal status (Figure 1.23).[1] However, large numbers of urban Africans lived elsewhere in the towns and cities, either in African freehold properties, or as tenants in backyards. The continued housing of domestic servants throughout the White suburbs constituted a major feature of most towns.

Furthermore, the central city areas which were relatively integrated were made subject to the provisions of the Slums Act of 1934. This was applied for the demolition of various inner but dilapidated suburbs, notably on the Witwatersrand (Parnell 1988). The displaced African populations were largely rehoused in segregated mono-racial municipal housing estates on the urban periphery. The Slums Act was widely applied in the major metropolitan centres in order to impose racial segregation in a 'non-racial' manner.

The development of state housing schemes after the First World War was also a means of enforcing segregation. Under the 1920 Housing Act central government funds were made available to local governments to build housing for the poor (Parnell 1989). Although the Act contained no racial connotations, the then Minister of Public Health, Sir Thomas Watt, upon introducing the Bill to parliament, indicated that he expected that local authorities 'will do their duty and provide for the coloured and native people within *their areas*' (emphasis added) (*Hansard* 1920: 185). The resultant housing estates were thus racially segregated, separated from one another by open spaces and with separate access roads. Although African municipal estates tended to be grouped in formal locations, those for the Coloured and White populations tended to be dispersed throughout the poorer parts of towns, without any concept of broad urban divisions. Figure 1.27 (see page 42) indicates the situation of such estates within the wider zoning practices of the private housing market.

The Indian community was also the target of various discriminatory measures from colonial times onwards. In the 1890s the Transvaal government sought to restrict the commercial activities of Indians. Separate 'bazaars' were established on the edges of existing towns where Indians were required to live and conduct their businesses. As in the case of African segregation measures, those applied to Indians were enforced with great variability,

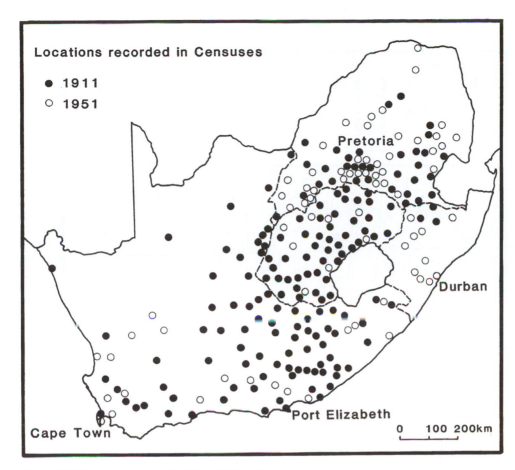

Locations recorded in Censuses

● 1911
○ 1951

Pretoria

Durban

Port Elizabeth

0   100 200km

Cape Town

Figure 1.23 Locations in South Africa, 1911–51
Source: Compiled from manuscript location returns in the 1911 and 1951 censuses, then held in the Central Archives Depot, Pretoria, and by the Central Statistical Services, Pretoria, respectively

with few bazaars effectively occupied before 1900. The Asiatic Bazaar in Pretoria housed only 60 per cent of the Indian population of the city (Figure 1.24). It also included a separate 'Cape Location' for the Coloured population. The issue of excluding Africans from such areas occupied the minds of the authorities until the 1930s (Parnell 1991). In Natal dual central business districts were created, with the Indian sector immediately adjacent to the White sector (Davies and Rajah 1965).

Anti-Indian agitation by the White population of Natal and the Transvaal resulted in an increasing level of restriction being placed upon the residential options of the community throughout the Union period. In 1922 the Natal Provincial Council passed an ordinance permitting the Durban Corporation to exclude Asians from accepted White suburbs. The policy was 'to separate the population of European descent, as far as possible, from Asiatics and Natives in residential areas – not to segregate any section or class entirely in parts of the Borough or from areas where at the present time any section has property or interests' (Durban 1923: 21). The attempt to introduce national legislation under the Class Areas Bill

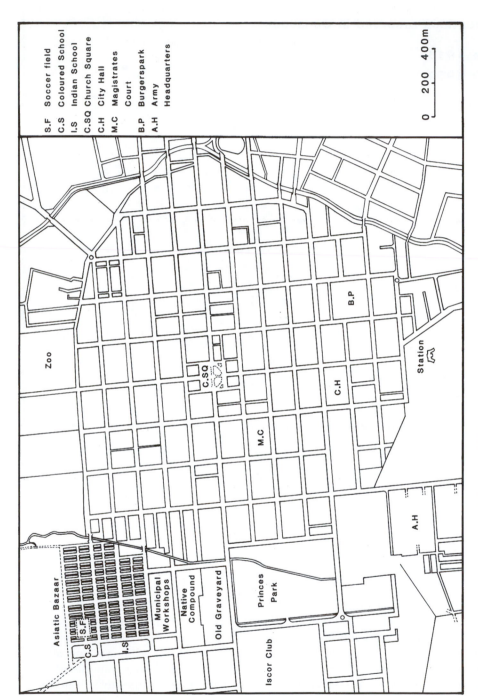

Figure 1.24 Asiatic Bazaar and central Pretoria, 1950

Source: Redrawn from the 1:10,000 map of the City of Pretoria (1950), Government Printer

of 1924 was withdrawn as a result of pressure from the Indian Imperial government in Delhi.

In the 1930s attempts were made in the Transvaal to place limits on the extent of Asian occupation, through a detailed investigation by the Transvaal Asiatic Land Tenure Commission. Asian rights to acquire land remained tightly circumscribed, although through the employment of dummy companies and front men it was possible to avoid some of the impact of the law. At the same time the bazaar areas were occupied and the provisions of the Housing Act applied to the Indian population.

White complaints of Indian 'penetration' into predominantly White residential areas in the 1940s in Durban resulted in further government action (Figure 1.25) (South Africa 1943). The enactment of the Trading and Occupation of Land (Transvaal and Natal) Act of 1943 and the more restrictive Asiatic Land Tenure and Representation Act of 1946 sought to confine Asian ownership and occupation of land to certain clearly defined areas of towns. This was to be achieved first through preventing inter-racial property transfers and

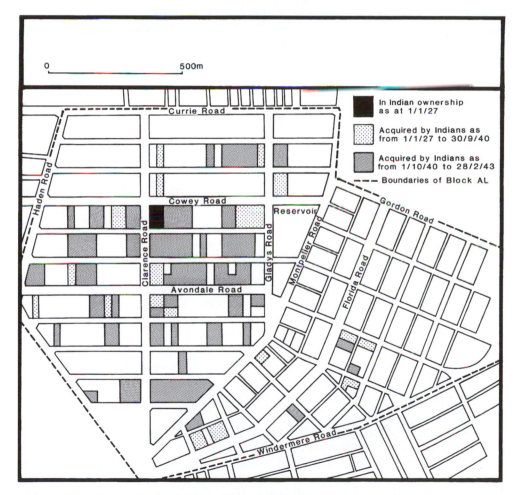

Figure 1.25 Indian purchase of property in central Durban
Source: After South Africa (1943) *Report of the Second Indian Penetration (Durban) Commission*, Pretoria: Government Printer

then by establishing a Land Tenure Advisory Board to draw up plans for a permanent division of the cities into White and Indian sectors (Maharaj 1995). The highly restricted range of Indian residence in central Durban is demonstrated by the 1951 census results (Figure 1.26).

In the private housing market, township developers and some individuals introduced racially restrictive covenants into the title deeds. Usually these were designed to prevent anyone who was not regarded as White from owning or leasing the properties. The practice was begun in Natal in the 1890s as an attempt to prevent Indians from acquiring properties in the more expensive, and now literally 'exclusive', suburbs. Strict injunctions were often inserted into deeds of transfer, specifying those who were not wanted. A Durban deed, for example, forbade the ownership or occupation of the land by any 'Arab, Malay, Chinaman, Coolie, Indian, Native or any other coloured person'.[2] However, private housing covenants included provision for resident African or Coloured servants, who were exempted from the restrictions: 'only as such necessary domestics as the master or mistress of the dwelling house may require for his or her purpose in the comfortable habitation of the said dwelling house'.[3] By the 1930s few new private residential developments were laid out anywhere in the country without such restrictions. The extent of restrictive covenants may be gauged from the situation in Port Elizabeth in 1950 where most of the suburbs developed in the twentieth century were restricted to the ownership and occupation of a single group (Figure 1.27).

The residential options open to Coloured and Indian people were therefore progressively restricted, although not to the same extent as those open to African people (Kuper *et al.* 1958). Segregation within the urban areas of South Africa was accordingly well advanced by the time the National Party came to power in 1948, and was at a similar level to that experienced by the Black population of the United States at the same time (Massey and Denton 1993). Only the Coloured and White communities in the western portion of the Cape Province experienced anything approaching an integrated residential pattern. Elsewhere racially defined segregation levels indicated the effects of long-enforced and entrenched coercive measures.

LEVELS OF URBAN SEGREGATION

Despite the range of segregationary legislation and legal restrictions in operation after 1910, distinct regional patterns of enforcement and effectiveness are evident. The colonial heritage of regional segregation practices remained until the impact of the era of apartheid in the 1950s (Christopher 1990). Thus the White population, for whom the whole exercise was designed, exhibited markedly different patterns of urban segregation in the different provinces at the time of the 1951 census. If the standard index of segregation, measured on a scale from 0 (completely integrated) to 100 (completely segregated), is examined, the populations of South African towns and cities were markedly segregated by 1951, often as a result of the legal structures outlined.[4]

The residential separation of the White urban populations was most noticeable in all four provinces (Figure 1.28). The median index of segregation for the White community registered 74.8. There were remarkable regional variations with significantly higher indices in the Eastern Cape and Orange Free State and significantly lower in Natal. These reflect the contrasting colonial heritages of the different regions of the country. An apparent anomaly in the figures is the low level of segregation in the Transvaal. In many respects this was due to the incorporation of extensive smallholding zones within the municipal boundaries, where

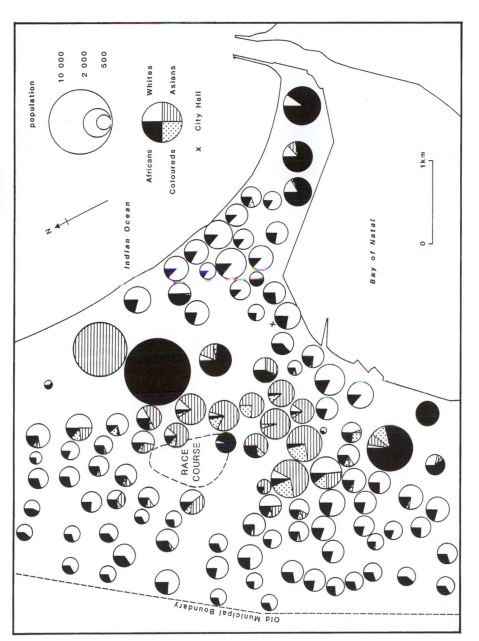

Figure 1.26 Distribution of population in central Durban, 1951
Source: Compiled by the author from the manuscript enumerators' returns of the 1951 census then held by the Central Statistical Services, Pretoria

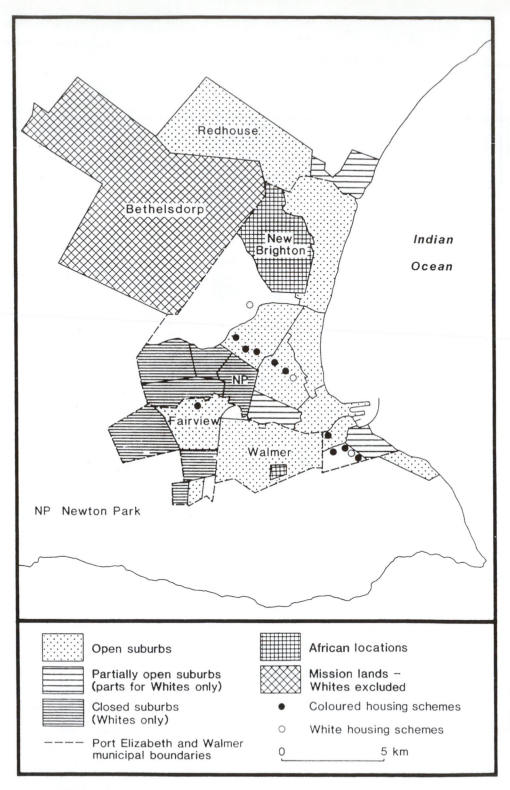

Legend:

- Open suburbs
- Partially open suburbs (parts for Whites only)
- Closed suburbs (Whites only)
- ---- Port Elizabeth and Walmer municipal boundaries
- African locations
- Mission lands – Whites excluded
- ● Coloured housing schemes
- ○ White housing schemes

0   5 km

Figure 1.27 Extent of zoning by title deed in Port Elizabeth
Source: Compiled by the author from Port Elizabeth municipal records and the Deeds Office, Cape Town

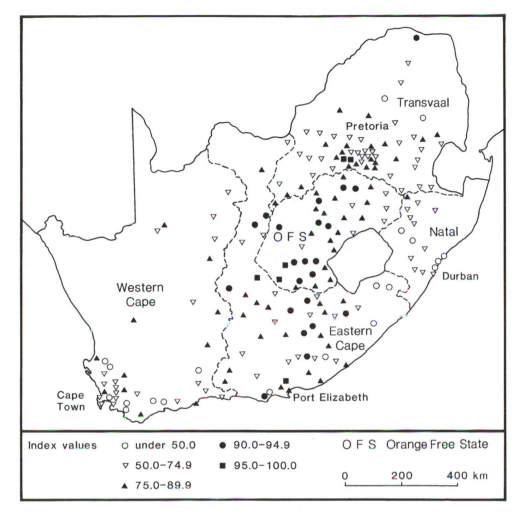

Index values   ○ under 50.0   ● 90.0–94.9   O F S   Orange Free State

▽ 50.0–74.9   ■ 95.0–100.0

▲ 75.0–89.9   0   200   400 km

Figure 1.28 White index of segregation, 1951
Source: Compiled by the author from the manuscript enumerators' returns of the 1951 census
then held by the Central Statistical Services, Pretoria

large African labour forces were housed on what amounted to small farms owned by White suburbanites. Again this continued the nineteenth-century colonial pattern where the majority of towns incorporated a substantial agricultural component.

The national level of Coloured segregation indices was substantially lower, with a median value of 49.3 in 1951 (Figure 1.29). Regional variations were again extremely marked. In the Orange Free State the Coloured population had generally been housed with the African community in the locations. On account of the numbers of Coloured domestic servants and labourers residing in the White areas, the community in the Orange Free State therefore surprisingly registered the lowest indices. In contrast, the small Coloured community in Natal had for most purposes been integrated with the White population. The majority of the Coloured population of the country lived in the Western Cape, where the majority of towns

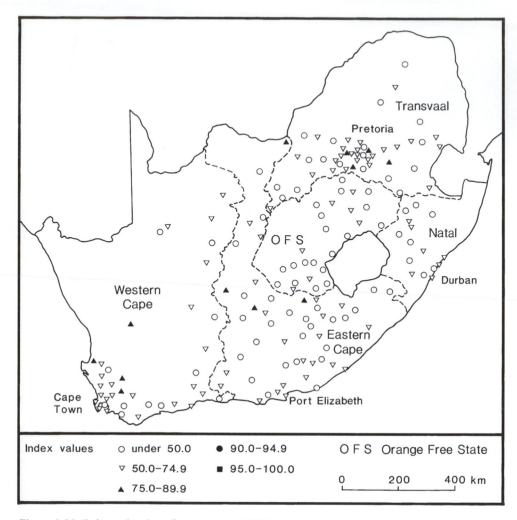

Figure 1.29 Coloured index of segregation, 1951
Source: Compiled by the author from the manuscript enumerators' returns of the 1951 census then held by the Central Statistical Services, Pretoria

exhibited only moderate levels of segregation with a median index of 55.8. In a number of studies of American cities an index value of 70.0 has been taken to indicate a measure of structural segregation (Massey and Denton 1989). In 1936 the median index had stood at only 44.3, indicative of the impact of segregated housing policies in the intervening fifteen years.

Asian levels of segregation exhibited a substantial variation in values (Figure 1.30). Even the high Transvaal median value of 80.1 masks a wide range from 32.5 to 96.6, reflecting the variety of municipal programmes directed towards the relocation of Indians in the bazaars and other segregation practices. Significantly, segregation levels between Asians and Africans were higher than those between Asians and Whites, as White urban administrations had made a concerted effort to exclude Asian retail stores from African areas and thus enable White traders to retain part of the African trade.

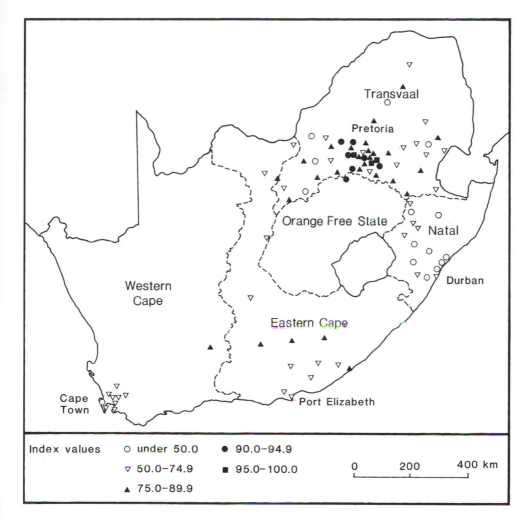

Figure 1.30 Asian index of segregation, 1951
Source: Compiled by the author from the manuscript enumerators' returns of the 1951 census then held by the Central Statistical Services, Pretoria

The African indices of segregation exhibited a range of values corresponding in many ways to the White indices (Figure 1.31). In the Eastern Cape and the Orange Free State high index values reflected the degree to which the African population had been confined to locations or compounds. In these two regions over 70 per cent had been so restricted, a proportion little changed over the previous forty years. In Natal the slow progress of location building was apparent as only a third of urban Africans lived in formal locations, compared with a national average of two-thirds. The remainder continued to live on their employers' properties, whether domestic or industrial.

It must be emphasized that in the pre-apartheid era the majority of urban dwellers lived in segregated circumstances. In most cases this was the result of structural segregation practices ranging from racially restrictive title deed covenants to legislative prohibitions. However, the

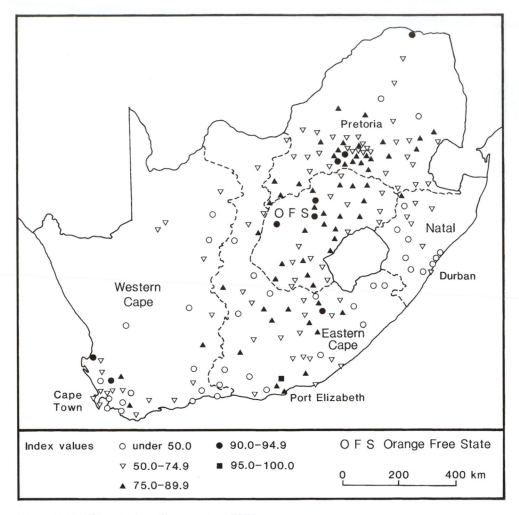

Figure 1.31 African index of segregation, 1951
Source: Compiled by the author from the manuscript enumerators' returns of the 1951 census then held by the Central Statistical Services, Pretoria

segregated suburbs were often separated by no more than a road, resulting in greater contact than the statistics might suggest.

THE SEGREGATION CITY

Pre-apartheid South African cities with their varying mixture of racial integration and enforced separation were described as 'segregation cities' (Figure 1.32) (Davies 1981). As such they reflected the complex heritage of British colonial cities, whose administrators had veered between attempts at enforcing systematic racial segregation and the pragmatism of accommodation and even selective integration (Christopher 1983, 1992). The results were

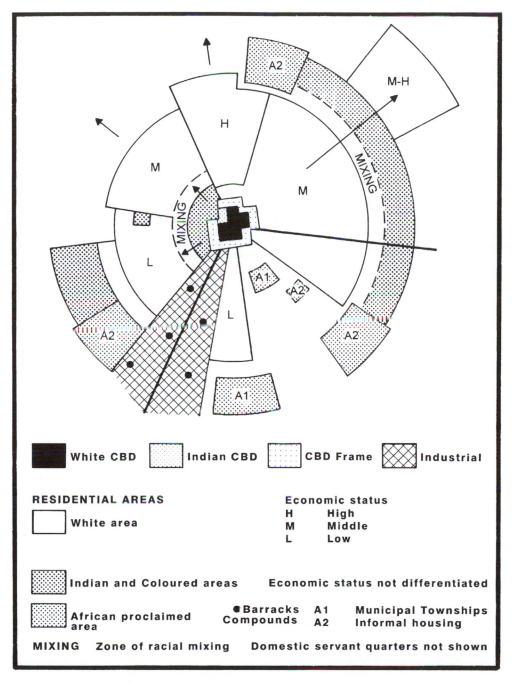

Figure 1.32 The model segregation city
Source: After R.J. Davies (1981) 'The spatial formation of the South African city', *GeoJournal* Supplementary Issue 2: 59–72

essentially untidy, reflecting the histories of the settlements concerned and the ethnic diversity of the urban population. There was usually a notable lack of any coherent plan, as administrations responded to crises as they arose (Christopher 1987; Davies 1963). Even the central business districts were contested between the European and Indian communities in the Transvaal and Natal, leading to the dual structures evident most notably in Durban. It was this inherent untidiness of the nineteenth-century city which conflicted with White South African ideas of town planning and racial order inherent in the state-driven programmes of reconstruction in the post-Second World War era (Mabin and Smit 1997).

2 Administering apartheid

The administrative machinery developed by the South African government between 1948 and 1991 to give effect to the apartheid policy was highly complex and expensive, involving the establishment of numerous overlapping jurisdictions and duplication of authorities. Some were inherited from the previous government, but the majority were new, and involved specifically with the implementation of the policy of apartheid. As a result, the bureaucracy expanded and by the early 1990s some 1.2 million civil servants were employed by the various administrations. Some understanding of the administrative pattern is therefore necessary in order to follow the process and account for the regional variations in the outcome of government policies.

The Union of South Africa was established through the unification of four British colonies in 1910. Under the constitution little power was retained by the four provinces, and a highly centralized state came into being. Only such concerns as school education, health, environmental protection and local roads remained under provincial control. Some regional peculiarities, such as the divisional council system in the Cape Province, survived until the 1980s with a measure of control by the local White electorate over social services in the predominantly rural areas. The various urban administrative systems provided for a greater measure of local control over municipal affairs. Even the spread of participation was wider as Coloured voters remained on the Cape municipal electoral rolls until 1972. Thereafter Coloured and Indian civic participation was limited to the introduction of Management Committees for the relevant group areas which could liaise with the White town council. A few Management Committees subsequently became fully fledged town councils, but this option was generally inoperable owing to the lack of financial resources and the Coloured and Indian communities' rejection of separation from the White municipality of which they had previously been a part.

Opposition to the central government's programme was thus only possible regionally from the provinces and the municipalities, where they were controlled by non–Nationalist councils. In the former case only the Natal Provincial Council was controlled by the opposition after 1948, although the Administrator was appointed by the central government. City and town councils were often ambiguous in their opposition to the government (Maharaj 1996a; Robinson 1996). Durban Corporation assisted in the drawing up of the guidelines for group area demarcation, yet was anti-Nationalist in other aspects of policy; the Cape Town City Council opposed all aspects of the implementation of Group Areas legislation as far as possible, although limited by the control which the central government exercised over local government finance (Western 1996). The era, though, was one of increasing levels of centralization where elected provincial councils were finally abolished and municipalities lost much of the independence that had previously existed. In place, a series of centrally controlled bodies ensured the dominance of the central government in carrying out the official programmes of the state.

The central government ministries maintained regional offices where necessary, staffed by professional bureaucrats who were transferred from one part of the country to another in the course of their careers. The opportunities for the development of regional bias were therefore weak. However, through the creation of pools of local expertise, local offices were able to modify the details of central government policies.

The most important local administrative unit of central government was the magisterial district. This was originally devised for the administration of justice and the collection of revenue in colonial times. Because of the significance of these functions, it was the maps showing magisterial districts which were most commonly used for administrative purposes and for the collation of statistics. The districts were based on urban spheres of influence until the 1950s, and generally disregarded the boundaries of African land holdings (Figure 2.1). The standard topographic maps depicted magisterial district boundaries and they were therefore prominent in the cartographic presentation of state policies. In contrast, municipal boundaries did not appear on the national map series, as they were subject to frequent change. However, district boundaries were not permanent and the constant creation of new

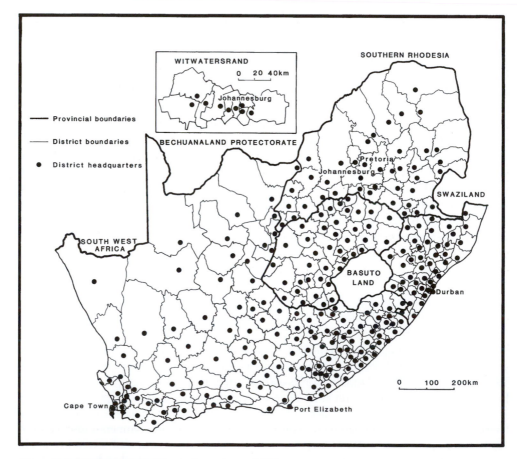

Figure 2.1 Magisterial districts and provinces, 1951
Source: After South Africa (1955) *Population Census 1951*, Pretoria: Government Printer

districts makes the comparison of statistics remarkably difficult at anything other than national level. By the time of the 1985 census, district boundaries reflected African homeland boundaries in a complex system of jurisdictions, which had substantially modified the map of magisterial districts in the eastern portion of the country (Figure 2.2). Even the boundaries of the four provinces had been redrawn in the 1970s and 1980s as a result of the implementation of the state partition policy.

AFRICAN ADMINISTRATION

Parallel with the civil administration was that responsible for the supervision of the African population (Evans 1997). The Department of Native Affairs and its successors operated a parallel administrative system through Native Commissioners and successor officers throughout the period from 1948 to 1991. The administrative apparatus combined with that of the magistracies in the areas covered by the Native Reserves, but was run parallel to it in the remainder of the country. In 1948 there were six Chief Native Commissioners (later eight) and a substantial subordinate hierarchy (Figure 2.3). Notably, the Chief Native

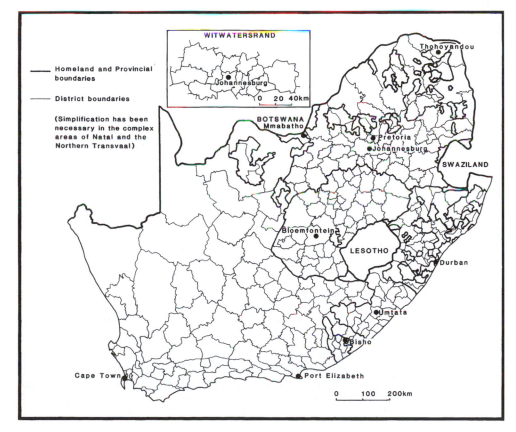

Figure 2.2 Magisterial districts and homelands, 1985
Source: Modified after South Africa (1986) *Population Census 1985*, Pretoria: Government Printer, and maps of the four independent homelands

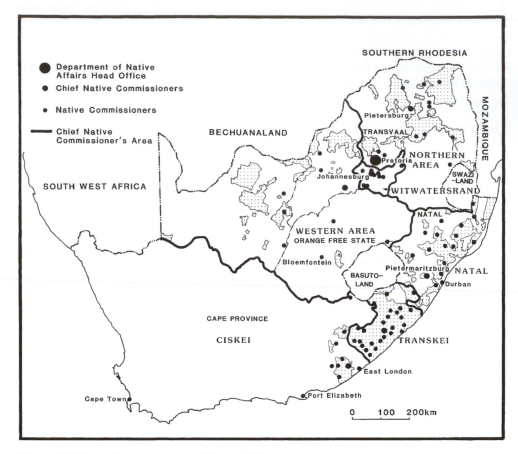

Figure 2.3 Native Commissioners' offices, 1955
Source: After South Africa (1955) *Summary of the Report of the Commission for the Socio-Economic Development of the Bantu Areas within the Union of South Africa*, Pretoria: Government Printer

Commissioner in Johannesburg also held the title Director of Native Labour, possibly an indication of the purposes of the entire regime. Regional offices of this department controlled most aspects of the lives of the African population, notably through the operation of the detested pass system until 1986. Passes amounted to internal passports allowing African persons to move and reside in specified parts of the country for specified periods of time. The acquisition of urban passes became progressively more difficult as the government imposed restrictions on permanent residence in the urban areas in favour of migrancy from the homelands. The later workings of this administrative system within the African rural areas will be dealt with in Chapter 3 on state partition. The administrative machinery was split between that applicable to the homelands and that responsible for the remainder of the country, under the Promotion of Bantu Self-government Act of 1959. In the former case Commissioners-General were appointed to oversee each of the African ethnic groups recognized in the new organization.

A word on terminology is required at this point. Frequent changes in nomenclature afflicted virtually all government functions. Nowhere was this more evident than in the realm of African administration. In 1955 the Department of Native Affairs was divided into the Department of Bantu Administration and Development and the Department of Bantu Education (subsequently the Department of Education and Training). The former was renamed the Department of Plural Relations and Development in 1976, then the Department of Co-operation and Development, only to dissolve into a host of departments in the 1980s, including the Department of Development Aid, as functions were transferred to the homeland authorities. The various parallel organizations, except those relating to education, were finally merged with national departments in 1991 as a part of the major reform programme introduced the previous year. The Department of Development Aid was racked by a series of financial scandals relating to resettlement programmes. Townships consisting solely of hundreds or thousands of lavatories, but no houses or people, pointed to massive misappropriation of the taxpayers' money. The 'toilet town' of Restaurant in Lebowa became the ultimate *cause célèbre* of a Department which lacked effective parliamentary supervision.

In the urban areas the separation of African administration from that of the remainder of the city was effected through a series of measures. The Native Advisory Boards set up under the 1923 Natives (Urban Areas) Act were replaced in 1961 by Urban Bantu Councils. These remained largely advisory bodies and acted as agents of the local (White) authorities. In 1971 control of local African affairs was transferred from the White local authorities to twenty-two Bantu Affairs Administration Boards, under the direct control of the Department of Bantu Administration and Development (Figure 2.4). After the 1976 Soweto riots the central government replaced the Urban Bantu Councils with Community Councils. Some 224 such councils had been established by 1980 but were unable to achieve any notable legitimacy among their nominal constituents. In 1982 the Black Local Authorities Act provided for the establishment of a series of local government structures similar to those operating in the White areas. In the 1983 local authority elections only 7 per cent of the potential African electorate cast votes for the new system, resulting in a significant problem of legitimacy. Thus many African urban areas continued to be administered by White officials appointed to the task following the boycott of elections or the subsequent resignation or even murder of collaborative African politicians.

One of the more bizarre features of the post-1982 administrative system was the creation of a host of new racially denominated municipalities with new names, leading to the dual naming of most towns (Figure 2.5). In many cases the names were those previously adopted for the African 'location', for example Soweto, formerly part of Johannesburg. In others the names were expressive of the area, for example Ibhayi, 'The Bay', at Port Elizabeth. In others the names were more overtly political, if highly inappropriate, as for example Thembalesizwe, 'Hope of the Nation', at Aberdeen (eastern Cape). In the majority of cases the new African or Coloured municipality housed a larger population than the White town, which retained the original name (Payne 1992). However, on the official survey map series of the country and indeed in most national and international atlases the name of the White town remained, often with no reference to the new municipalities.

REGIONAL PLANNING

Regionalization was a constant theme as the central government sought to streamline the increasing complexity of local government structures (de Villiers 1981). In 1982 the Department of Planning instituted a series of eight (later nine) planning regions for the

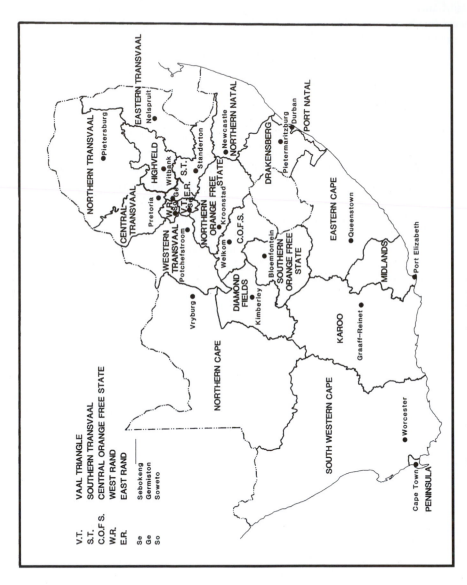

Figure 2.4 Bantu Affairs Administration Boards, 1972–3
Source: Based on information in the *Government Gazette*, various dates 1972–3

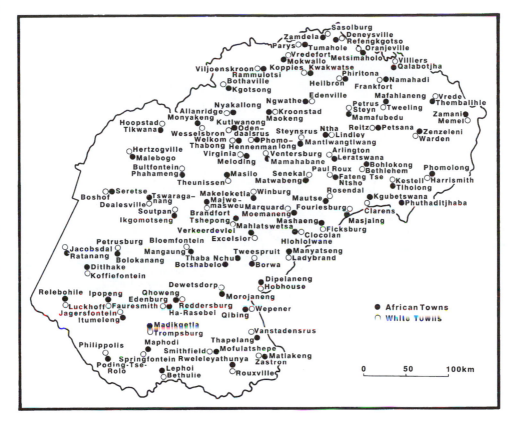

Figure 2.5 Town names in the Orange Free State
Source: Modified after R. Payne (1992) *The Two South Africas: A People's Geography,* Johannesburg: Human Rights Commission, and Orange Free State directories and maps

country to co-ordinate national planning, notably that related to economic development (Figure 2.6). The regions were based on the major metropolitan regions, and incorporated the various African homelands. This was in marked contrast to the majority of earlier national plans which had specifically excluded them. The move constituted one of the first major departures from the policy of state partition. Significantly, some homelands, notably Transkei, were divided between regions.

In 1985 the Regional Services Council Act provided for the establishment of joint metropolitan authorities for the distribution of bulk services (Seethal 1991). The first councils were established in 1987 in the major metropolitan areas and extended the following year to the rural areas. The councils were indirectly elected by the various authorities represented on the council. However, the voting strength bore no relationship to population numbers, but was based on the financial contributions paid for the services consumed. Thus the Central Rand Regional Services Council had fourteen participating bodies, four African, two Coloured, four Indian and four White. The voting strength of the White authorities totalled 74.5 per cent, while the four African authorities constituted 19.8 per cent, although in population terms the proportions were reversed (Lemon 1992). The Regional Services Councils were thus constituted in such a manner as to ensure White

Figure 2.6 Planning regions, 1982
Source: Based on the Development Bank of Southern Africa (1985) *Map of South Africa*

control of this intermediate level of local government. A major exception to the Regional Services Council system was the Natal region where the KwaZulu government opposed its introduction, preferring a more equitable sharing of power and resources between African and White areas through the inclusion of homeland local authorities. In 1991 the government relented and Joint Services Boards were established for the Natal-KwaZulu region (Figure 2.7).

CONTROL OF PARLIAMENT

The point has been made that, although the franchise was highly restricted, South Africa was a parliamentary state and the control of parliament through elections was essential for the implementation of policy (Welsh 1998). Indeed it was this degree of parliamentary responsibility which resulted in the wealth of reports from commissions of inquiry, annual departmental reports and detailed answers to questions raised by the

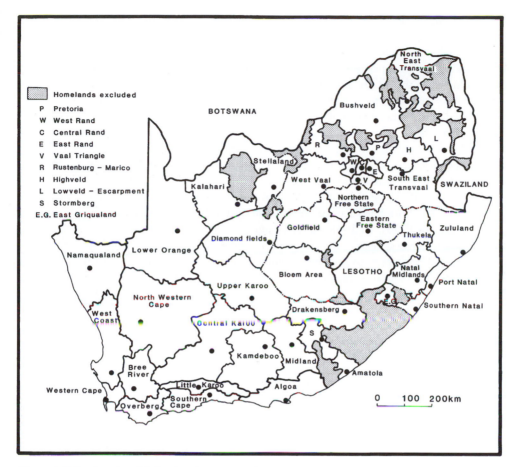

Figure 2.7 Regional and Joint Services Councils
Source: Based on information supplied by the four provincial administrations

members of parliament which provide so much information about the era. There were limits to how far the government could act in secrecy: although in restricted spheres such as nuclear arms and death squads it could do so, in general its activities were documented and publicized.

Before 1994, control of parliament was exercised through political parties, which were dependent upon White Afrikaner support. Afrikaner nationalism was effectively mobilized by Dr D.F. Malan and his Herenigde (Reunited) National Party against the perceived imposition of British ideas and the threat of racial integration by the ruling United Party. Part of the effectiveness of the Afrikaner national movement was its high degree of organization and dedication. The Afrikaner Broederbond (Brotherhood), founded in 1918, played a significant role in generating an appreciation of the Afrikaans language and mobilizing an awareness of an Afrikaner identity and separate nationhood (Wilkins and Strydom 1978). The membership was small and elite. It operated as a secret society furthering its ideals and the individual promotion of its members, who were White Afrikaner males over the age of 25. The Broederbond was strongly represented in key sectors of public influence, notably in commerce, industry, the Afrikaans press, education, the Afrikaans churches, the army and the state bureaucracy. Notably, all the Nationalist prime ministers and presidents were members.

Membership of the Broederbond was acquired by invitation and was always highly select-ive. Thus fewer than 1 per cent of eligible Afrikaner men were actual members at the height of its influence in the 1970s. The proportion forming the membership was also geographic-ally uneven, according to a study undertaken in the late 1970s (Pirie *et al.* 1980). The dominance of the rural areas, notably the southern Orange Free State, north-eastern Cape and northern Natal, is most noticeable (Figure 2.8).[1] In contrast, only the administrative capital Pretoria emerges with above anticipated membership within the major urban centres. The industrial heartland, the Witwatersrand conurbation, appears to have been relatively unaffected, as were the major port cities.

The most significant feature of the political system was the monopoly of political power exercised by the National Party between 1948 and 1994 in the aptly labelled 'forty lost years' (O'Meara 1996). In 1948 the Herenigde National Party and its ally the Afrikaner Party won a majority of seats in the House of Assembly, although only a minority of the vote (Figure 2.9). The element of gerrymandering within the delimitation process was such that 40 per cent of the vote was translated into 53 per cent of the seats, even though the ruling United Party had controlled the process and inadvertently assigned to itself larger than average constituencies and its opponents smaller than average, with no appreciation of the

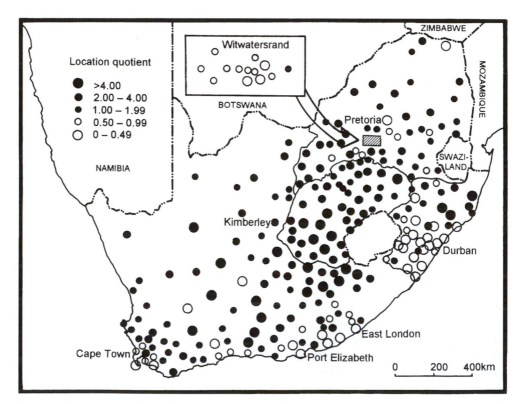

Figure 2.8 Relative strength of Broederbond membership
Source: After G.H. Pirie, C.M. Rogerson and K.S.O. Beavon (1980) 'Covert power in South Africa: the geography of the Afrikaner Broederbond', *Area* 12: 97–104

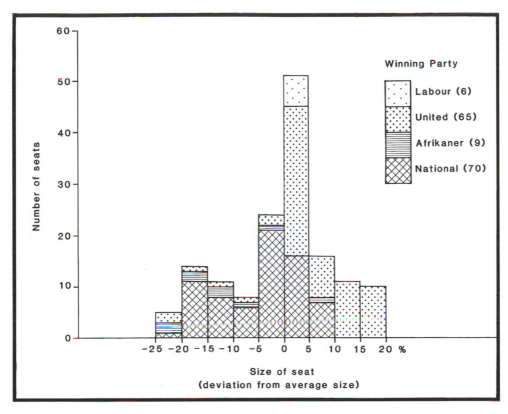

Figure 2.9 Size of constituency and winning party, 1948
Source: Based on constituency sizes in *Government Gazette* No. 3931, 13 February 1948, and election returns in the *Eastern Province Herald* and *Oosterlig*, 27 May 1948

consequences. In 1951 the Herenigde National Party and the Afrikaner Party merged to form the National Party.

The delimitation process was then manipulated to ensure that rural voters retained a disproportionate voice in parliament (Figure 2.10). Until the 1980s the National Party was always assured of the majority of rural votes. The Union constitution provided for a 15 per cent range on either side of the provincial average constituency size. In 1963 it became possible for rural constituencies to be created which were 30 per cent smaller than the average, thereby packing the urban vote yet further. Such options were taken as injunctions to gerrymander by successive delimitation commissions. The result was to boost National Party representation in the House of Assembly. Thus in 1953 the party, though only securing an estimated 45 per cent of the votes, obtained 59 per cent of the seats, while in 1966, the party's 59 per cent of the vote was translated into 75 per cent of the seats (Heard 1974). The large number of unopposed seats resulting from the regional concentration of voting support made any discussion of the electoral system in South Africa remarkably speculative. Indeed the extent of gerrymandering was such as to elicit the comment: 'The successive Delimitation Commissioners have produced an electoral framework, whether knowingly or not, which makes a mockery of democracy' (Gudgin and Taylor 1979: 140).

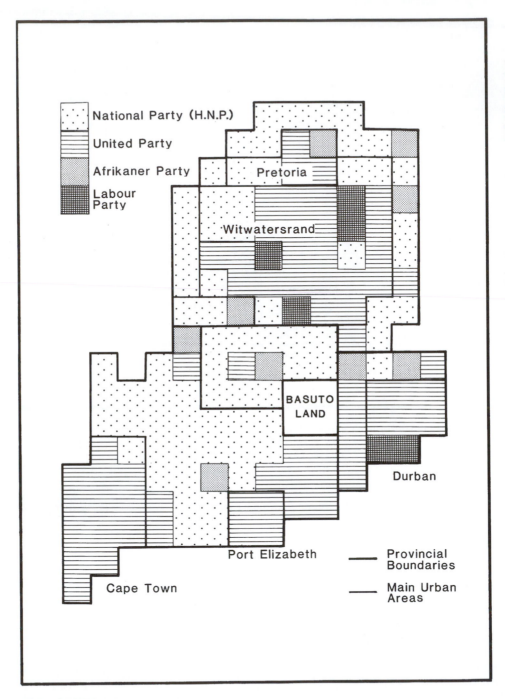

Figure 2.10 Election results, 1948
Source: Based on election returns in the *Eastern Province Herald* and *Oosterlig*, 27 May 1948

National Party support in the urban areas increased as Afrikaans-speaking Whites migrated to the towns. The government was also able to attract English-speaking support as a result of its racially based policies. Further, the opposition United Party slowly disintegrated as prospects of regaining power receded. It finally collapsed in 1977, enabling the National Party to obtain an estimated 68 per cent of the vote and 82 per cent of the seats in that year's election (Figure 2.11). Thereafter opposition from the Progressive Federal Party and its successor, the Democratic Party, on the left, presented no threat to the continued supremacy of the National Party in the House of Assembly. However, the Conservative Party, founded to oppose the President's reform programme in 1982, grew to became the official opposition five years later. The 1989 general election resulted in further government losses as the Conservative Party made significant inroads into the areas of traditional National Party support, notably in the rural districts of the Transvaal and Orange Free State (Figure 2.12) (Lemon 1990).

Parliamentary representation was also manipulated through the exclusion of the Coloured voters from the common voters' roll in the Cape Province in 1956. Exclusion only came after a five-year constitutional battle culminating in the packing of the Senate by the National Party to ensure the overruling of the entrenched constitutional safeguards concerning the franchise. At this time the Women's Defence of the Constitution League was founded, later to gain recognition for its human rights work as the Black Sash (Spink 1991). The number of Coloured voters was comparatively small, but highly localized and hence influential in marginal seats, which the National Party hoped to gain (Figure 2.13). The separate representation of the Coloured community by four (White) members of parliament was an interim arrangement before their total exclusion from parliament in 1968. The powerless Coloured Persons Representative Council was instituted as a substitute. This body was abolished after the Labour Party gained control of it in 1975 with the intention of making it unworkable.

In 1984 a new constitution was introduced with a tri-cameral parliament. A House of Representatives for the Coloured community, and a House of Delegates for the Indian community were added to the now all White House of Assembly to constitute the parliament (Lemon 1984). It should be noted that the degree of gerrymandering in the delimitation process for the two new houses was particularly blatant. In the case of the House of Representatives the number of registered voters in the constituencies ranged from 3,000 to 16,000, as a result of the differential weighting of the four provinces (Figure 2.14). The bias against the metropolitan areas of Cape Town and Port Elizabeth, which constituted significant centres of political opposition, is marked.

The balance of power between the three houses was heavily weighted in favour of the House of Assembly. The President's Council was established as an advisory body from the majority parties in the three houses, in which the ratio of members was 4:2:1 (Assembly:Representatives:Delegates). The President's Council was given the power to override the opposition of one or two houses of parliament to a piece of legislation, thus ensuring that the House of Assembly, dominated by the National Party, remained in ultimate control. Furthermore, the abolition of the post of Prime Minister and the institution of an executive presidency effectively concentrated political power in the hands of the President and his chosen inner cabinet and the state security apparatus, thereby reducing the importance of parliament.

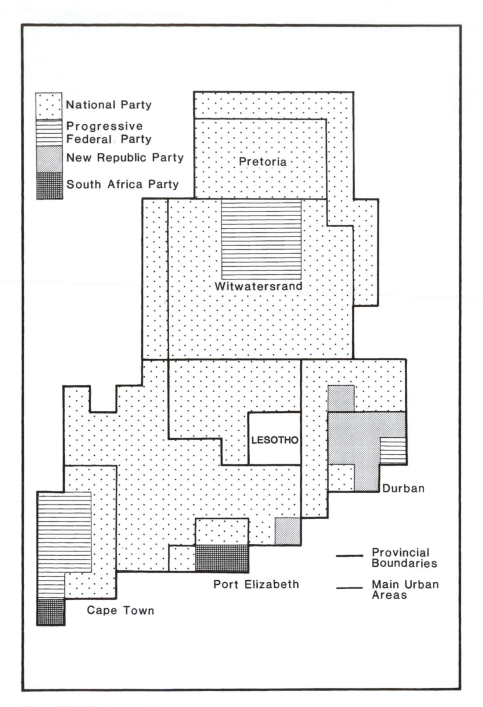

Figure 2.11 Election results, 1977
Source: Based on election returns in the *Eastern Province Herald* and *Oosterlig,* 1 December 1977

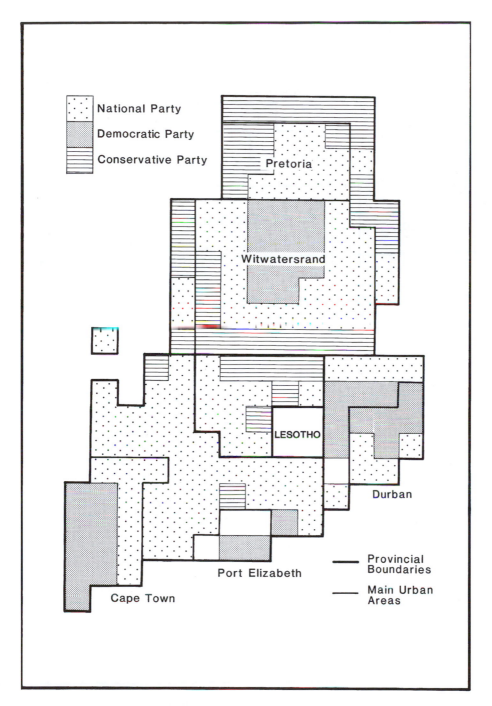

Figure 2.12 Election results, 1989
Source: Based on election returns in the *Eastern Province Herald* and *Oosterlig*, 7 September 1989

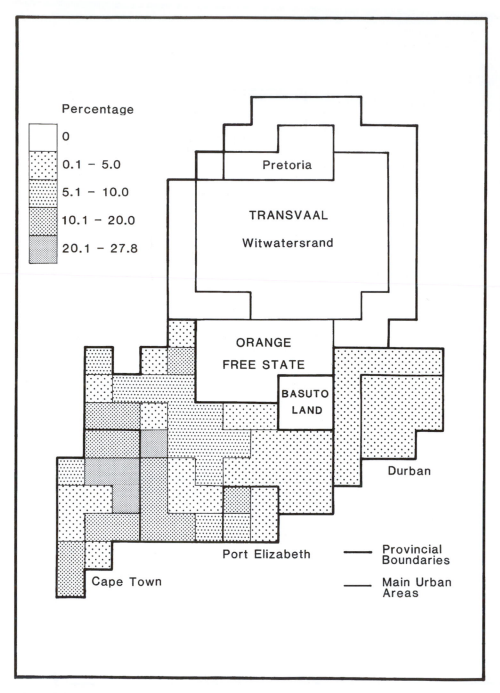

Figure 2.13 Percentage of electorate classified non-European, 1950
Source: Based on statistics in South Africa (1953) *Official Yearbook of the Union of South Africa 1950*, Pretoria: Government Printer

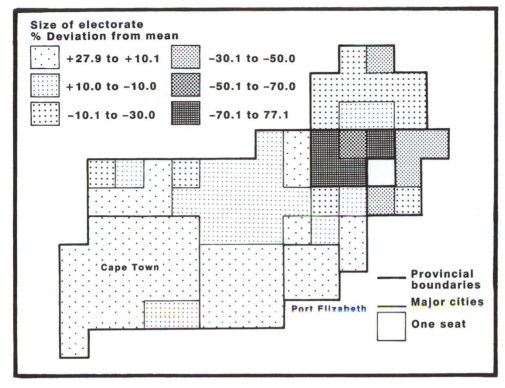

Figure 2.14 Gerrymandering the House of Representatives
Source: Based on statistics in Report of the Delimitation Commission for the House of Representatives and House of Delegates, *Government Gazette* No. 9253, 11 June 1984

ADMINISTRATIVE FRAGMENTATION

One of the features of the various constitutional changes in South Africa between 1948
and 1991 was the fragmentation of functions and jurisdictions, and the consequent multiplicity of overlapping authorities. Undoubtedly this provided a wide range of positions
within the civil service and hence the creation of a substantial interest group favouring the
continuation of the system (Crankshaw 1997). Thus for example, by 1991 there were
seventeen separate departments responsible for state school education, namely the ten
homeland authorities, the four provincial White authorities, those run by the House of
Representatives, the House of Delegates, and the Department of Education and Training
(which was responsible for the provision of African education outside the homelands).
Accordingly, in most towns and cities there were at least three if not more education
authorities responsible for the state school system. Although mother-tongue education
was only enforced in junior primary schools in the African education system, in the White
system it was pursued throughout. Thus in Pietermaritzburg, as in most other towns and
cities, the secondary system was even more confused, particularly as private schools were
largely denominational (Figure 2.15). The conflicting interests of the various education
departments inevitably led to crises in the implementation of Christian national education

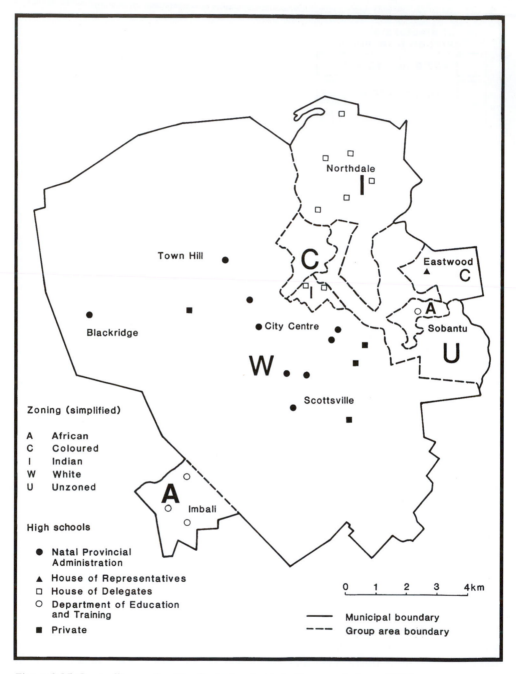

Figure 2.15 Controlling authorities for high schools in Pietermaritzburg, 1991
Source: Based on information supplied by the relevant education authorities

and Bantu education. Crisis was the constant state of affairs in the later years of the apartheid era as African school children formed one of the more visible and militant groups opposed to the political system. Similar duplication was evident in many other areas of social service administration.

3   State apartheid

The philosophy of apartheid was introduced as the guiding principle of the National Party government under Dr D.F. Malan, which came to power in the general election of 26 May 1948, with the intention of creating a White Christian national state. Initially the new government differed only in degree of enforcement from the segregationism practised by the previous administration. However, the guidelines were set in terms of the apartheid philosophy, which were to focus attention on South Africa in international affairs and give it its distinguishing features. Separation was to be introduced at all inter–personal levels ranging from separate park benches for Whites and other people to separate independent states for members of the various defined population groups. The principle of no equality between Black and White in church and state had been written into the Afrikaner republican constitutions of the South African Republic and Orange Free State in the nineteenth century, and the new government sought to emulate its predecessors in the mid-twentieth century.

State partition became the declared aim of the government as the policy unfolded. The major impetus to this policy came from Dr H.F. Verwoerd, who was appointed Minister of Native Affairs in 1950. In 1927 the Natives Administration Act effectively made the proclaimed African areas subject to a separate political regime from the remainder of the country, ultimately subject only to rule by proclamation, not parliament. The Ministry of Native Affairs, as subsequently renamed several times, controlled a wide range of policies through what amounted to a parallel administration run by Native Commissioners. Dr Verwoerd used these powers to establish tighter control over the African people both within what was perceived as the White part of South Africa and in the various Native Reserves and Trust Lands. Separate government for the African population was evolved into separate administrations, where the demands for political rights could be met without endangering White control over the remainder of the country. Thus state partition evolved as the cornerstone of apartheid to ensure continued White political control of South Africa in an era of African emancipation on the remainder of the continent. It should be noted that it was only under the pressures of a changing international political environment that Dr Verwoerd accepted the full implications of state partition as the solution to White South Africans' numerical inferiority (Pelzer 1966).

Under this policy all Africans in South Africa would become members of an African nation, which would possess a separate territorially based state, and within which the nation's political rights would be exclusively exercised. Thus at the end of the policy's implementation there would be no African South Africans; Whites, as the largest group of citizens, would be able numerically to continue to dominate the government of the rump state. The possibilities of creating states for the Coloured and Indian communities were periodically discussed, but their establishment never became official policy. It was the extinction of the African political presence in the country which was uppermost in the White policy makers' minds, as it was realized that African numerical superiority constituted the greatest threat to continued White control. In practical terms this perception was also to be translated into a programme for the removal of as many Africans from the White zone of South Africa as possible, leaving behind only those considered to be essential for the running of the country.

NEW NATIONS AS THE BASIS OF PARTITION

The first problem which the government confronted was the need to create African nations as the basis of nation-states. The government did not envisage the division of the country into African and non-African halves, where the African zone would be numerically and economically powerful. In a classic example of the principle of 'divide and rule', the African population was fragmented into a series of linguistically defined groups, which were the basic building blocks of the proposed dispensation. The broad African or Bantu linguistic families were fragmented into ten subdivisions or 'national units' (Figure 3.1). Thus the Sotho linguistic family was divided into North, South and West Sotho, or Pedi, Basutho and Tswana respectively. Similarly the Nguni linguistic family was divided into Xhosa, Zulu, Swazi, Shangaan, North Ndebele and South Ndebele. To these was added the Vendano of the northern Transvaal. It should be noted that in the 1950s none of these groups was significantly larger than the total White population, which was conveniently regarded as a unity, despite its linguistic and cultural diversity and political antagonisms.

TERRITORIES FOR NATIONS

The broad geographical distribution of these 'national units' gave some degree of order to the next stage of development. In 1959 the government enacted the Promotion of Bantu Self-government Act, which sought to create a hierarchy of local governments for the African rural reserves. Thus headmenships, chieftaincies, paramount chieftaincies and territorial authorities assumed a place in an orderly progression of power in the African areas. The African population was accordingly assigned what was officially regarded as the group's 'traditional territory', even though the distribution reflected only the nineteenth- and early twentieth-century reserves.

The territorial basis of the system was the land designated as Native Reserves and that belonging to the South African Native Trust. It needs to be remembered that the reserve boundaries were drawn according to the priorities of the colonial and other state governments, which certainly envisaged no form of separate sovereign existence for them. They were conceived as labour pools or homes for Africans surplus to White needs, but within the context of the South African political arena. They were remarkably fragmented with few consolidated blocks of scheduled land to provide any respectable territorial basis for the partition. The report of the Tomlinson Commission, which laid the framework for social and economic development in the 1950s, prompted the South African government to make further attempts to incorporate Basutoland, Swaziland and the Bechuanaland Protectorate into the Union (South Africa 1955b). They would then have formed the basis of three African states to disguise the small proportion (13 per cent) of the country being apportioned to the new African state structures (Figure 3.2).

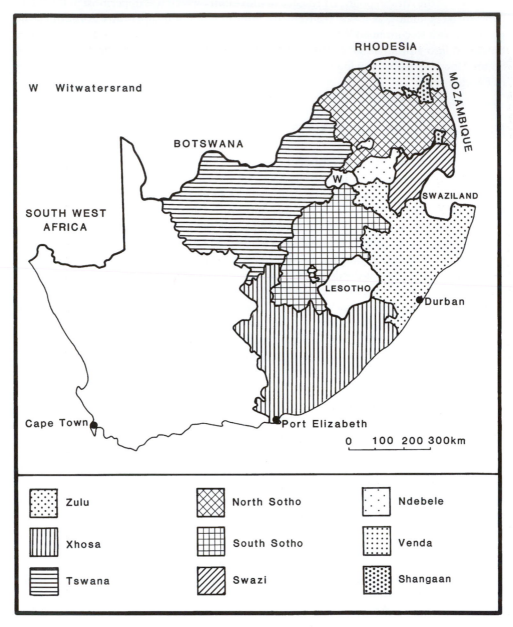

Figure 3.1 Ethnic divisions of the African population, 1970
Source: Based on statistics in South Africa (1973) *Population Census 1970*, Pretoria: Government Printer

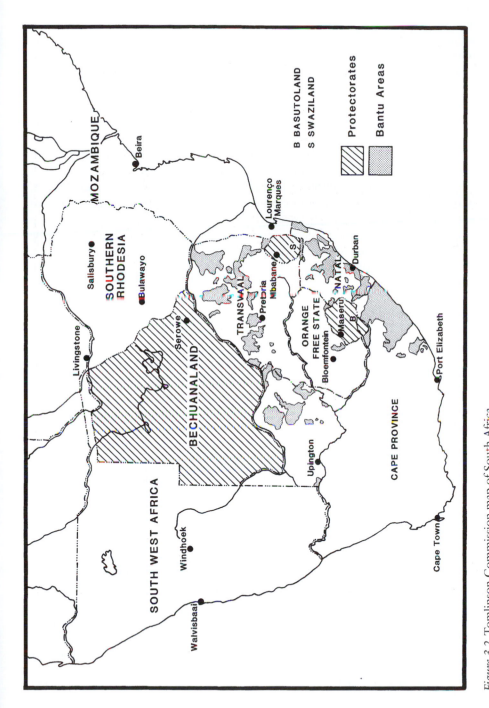

Figure 3.2 Tomlinson Commission map of South Africa
Source: After South Africa (1955) *Summary of the Report of the Commission for the Socio-Economic Development of the Bantu Areas within the Union of South Africa*, Pretoria: Government Printer

The various Native Reserves and blocks of South African Native Trust land were grouped together on a linguistic basis and brought under the supervision of a Chief Commissioner. The exception to this was the Xhosa people who were divided on historic grounds by the Kei River. Thus two Xhosa administrations evolved, one for the Ciskei and the other for the Transkei. Only in the latter had there been any system of consolidated local self-government before 1948. In 1966 the Ciskei government investigated the question of union with the Transkei, but rejected the option fearing loss of power to its larger neighbour and proceeded to foster a separate Ciskeian nationality (Pieres 1995). In the period 1986–7 the government pursued Operation Katzen, aimed at overthrowing the Ciskei and Transkei governments which covertly supported the liberation movements, and creating a united Xhosaland with a more favourably disposed administration. The planned *coups d'état* failed.

The other exception was the lack of any territorial base for the two smallest incipient nations, the North and South Ndebele, until the 1970s. The administrative pattern which emerged was one of considerable intricacy. The Zulu lands, for example, were divided into over a hundred separate blocks when privately owned lands were included, suggesting comparison with the medieval German state system (Figure 3.3).

A second problem was the small proportion of the African population living in the new states. In 1951 over 60 per cent of the African population of South Africa lived in the areas designated as forming part of the future 'White South Africa'. At state level, in virtually all cases, little more than half of those claimed to be citizens of the new states actually lived within their borders. As late as 1970 only 1.2 per cent of the South Sotho people of South Africa lived in the designated homeland of Qwaqwa, while 66.9 per cent of Vendans resided in Venda (Figure 3.4). Indeed, in that year only 41.8 per cent of the African population of the country lived in its designated state.

An added problem was the presence of ethnic minorities in virtually all the new states. Thus in 1970 only some 31.8 per cent of the Shangaan population of South Africa lived in Gazankulu, but another 21.5 per cent lived in other African homeland states. By judicial movements of population between the homelands and the incorporation of strategic territory some 43.3 per cent of the Shangaan population of the country lived in Gazankulu in 1985. Minorities remained a source of friction between several of the new state governments, particularly where the basic ethnic identity of the population was disputed. Census enumerators in the African states tended to find the characteristics of the dominant group rather than the minority, while compiling the census data, thereby statistically reducing the extent of the minority problem. Homeland governments exerted pressures on minorities to conform to the language and culture of the governing group or face the consequences of dispossession and expulsion (Vail 1991). In the process of nation building the power of the new states became increasingly important in the quest for conformity; and, resistance to that power often erupted in violent discontent.

The establishment of an Ndebele state illustrates the problems of a lack of a 'traditional territory' for the group. Until 1975 the South Ndebele constituted a minority within portions of Bophuthatswana and Lebowa, when they were encouraged to secede and form a separate state. The complexity of the situation in KwaNdebele was such that official plans were thwarted by the Bophuthatswana and Lebowa governments, which successfully sought to retain their territory in the region and expel the Ndebele inhabitants. Thus most of the state area had to be created from land previously in White hands, which was then populated by Ndebele people displaced from other areas, notably Bophuthatswana (Figure 3.5). Furthermore, the attempt by the KwaNdebele government to opt for independence in the mid-1980s was thwarted by violent internal dissension among a population consisting

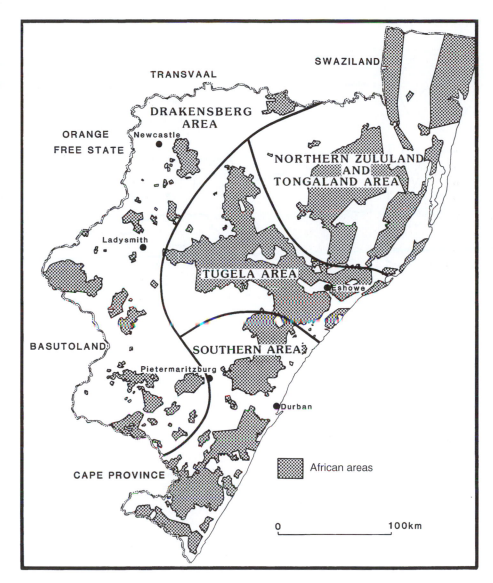

Figure 3.3 Fragmentation of Zulu lands in Natal, 1955
Source: After South Africa (1955) *Summary of the Report of the Commission for the Socio-Economic Development of the Bantu Areas within the Union of South Africa*, Pretoria: Government Printer

almost entirely of persons forcibly resettled from other parts of the Transvaal, including Bophuthatswana and Lebowa (Ritchten 1989).

CONSOLIDATION

Territorial consolidation of the scattered reserves was proposed in the Tomlinson Commission Report in 1955. The suggested plans involved the three High Commission Territories

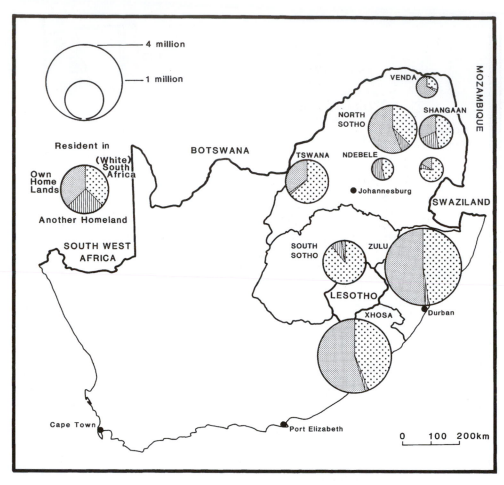

Figure 3.4 Proportion of African group in homelands, 1970
Source: Based on statistics in South Africa (1973) *Population Census 1970*, Pretoria: Government
Printer

and a remarkably radical redrawing of the map of South Africa, as the then existing African
areas were considered to offer 'no foundation for community growth' (Figure 3.6). The
government rejected such expensive plans and proceeded with the consolidation of the
reserves and Trust Lands into larger blocks through a series of judicious land exchanges and
purchases. Isolated tracts of African land were excised from the homelands, while strategically placed land was purchased to join other detached portions to form larger blocks. In this
manner the authorities expected to consolidate the African territories into at most six portions per state. A severe limitation upon the programme was the political undertaking that
the total area of the new states would not exceed the area specified in the 1936 Native Trust
and Land Act, namely 17 million hectares. It was only in the 1970s that this limit was
reached, after forty years of restricted implementation, and the undertaking was then
ignored. Clearly no broad, sweeping consolidation was envisaged in official circles in the
1930s to create a viable territorial base for the new states.

Several consolidation plans were produced, each one more comprehensive than the last

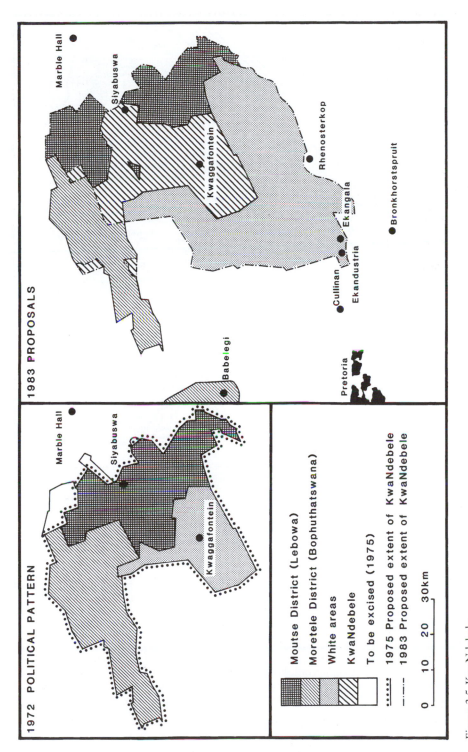

Figure 3.5 KwaNdebele
Source: Modified after L. Platzky and C. Walker (1985) *The Surplus People: Forced Removals in South Africa*, Johannesburg: Ravan

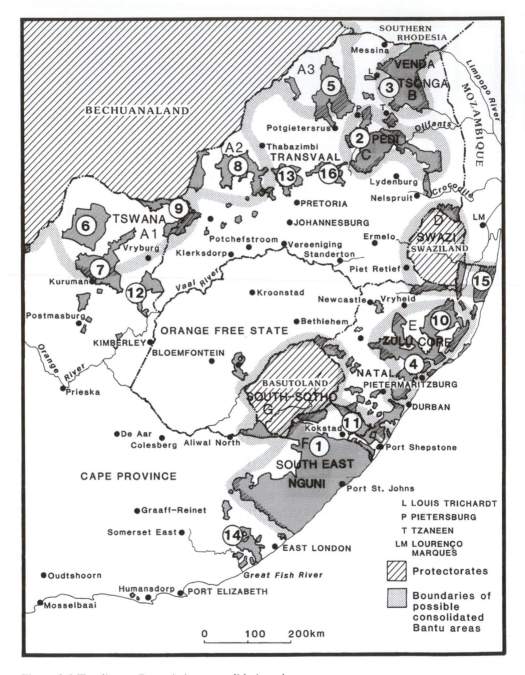

Figure 3.6 Tomlinson Commission consolidation plans
Source: After South Africa (1955) *Summary of the Report of the Commission for the Socio-Economic Development of the Bantu Areas within the Union of South Africa*, Pretoria: Government Printer

(Lemon 1976). The first comprehensive plan in 1973 involved a radical redrawing of the map of the African areas of South Africa and a structuring of the new states (Figure 3.7). Attention was directed towards those areas where most progress had been made with the political programme of self-government and where African co-operation was most evident and hence rewarded. This was reflected in the definitive 1975 plans which remained the basic framework until the demise of the policy (Figure 3.8). Thus the co–operative governments of Transkei and fragmented Ciskei received more attention than KwaZulu, where the African authorities resisted government policy. In part this also reflected the initial adherence to the provincial distribution of areas authorized for incorporation under the 1936 legislation. There was no outstanding balance available in Natal. A further complication was the presence of over 800,000 hectares of African privately owned land, which was also brought into the process of computing state areas. Accordingly, in Natal the scattered African private farms were progressively expropriated and compensatory land adjacent to the reserves acquired by the Bantu Development Trust. It should be noted that those farmers whose land was expropriated were entitled to no personal claim on the compensatory land from the homeland governments.

In the process of consolidation many of the smaller reserves and private farms were eliminated. These areas were referred to as 'black spots' to be expunged from the map. In 1961 it was estimated that there were approximately 330 farms with a total area of 148,000 hectares falling into this category. The substantial change in the outlines of Bophuthatswana is a case in point (Moolman 1977). The extensive unrestricted purchase of land in the Transvaal before 1913 and the Cape Province until 1936 had resulted in the emergence of a scatter of African privately owned lands in the south-western Transvaal and northern Cape (Figure 3.9).

The abolition of Native Reserve status frequently did not involve any form of compensation for the residents. The fate of the Mfengu Reserves in the Tsitsikamma Forest, some 150 kilometres west of Port Elizabeth, is an example where minimal compensation was paid for the houses, barns, etc., but none was offered for the land. The blocks of land had been granted by the Crown to be held in trust for the various bands of Mfengu (Fingos) who had been displaced during the colonial frontier war of the early 1830s (Figure 3.10). In 1977–8 the community was forcibly moved 350 kilometres away to Keiskammahoek in Ciskei. Its members did not receive compensatory land as individuals had not held freehold title deeds, only tickets of occupation, to the land they worked. The area involved was included in the general compensatory land transferred to the Ciskei government, not to the community involved. The South African government subsequently divided the Tsitsikamma land into nineteen farms which were then sold to White farmers. The demand by the Mfengu for restitution of the land, as indeed by other communities forcibly removed in the apartheid era, was a constant source of rural opposition to the policy.

The Tomlinson Commission had recommended the consolidation of the African areas through the incorporation of adjacent South African African-occupied territory into the three High Commission Territories. In 1982 the scheme was revived briefly with the attempt to transfer the Swazi national area (KaNgwane) and the district of Ingwavuma in northern KwaZulu to Swaziland (Figure 3.11) (Esterhuysen 1982). The latter piece of territory would have provided Swaziland with a coastline, while the former would have partially overcome Swaziland's claims for the restitution of territory lost to White colonists in the late nineteenth century. The scheme was undermined by the opposition of the KaNgwane and KwaZulu governments (Griffiths and Funnell 1991). The rectification of international boundaries which the plan involved incurred the opposition of the

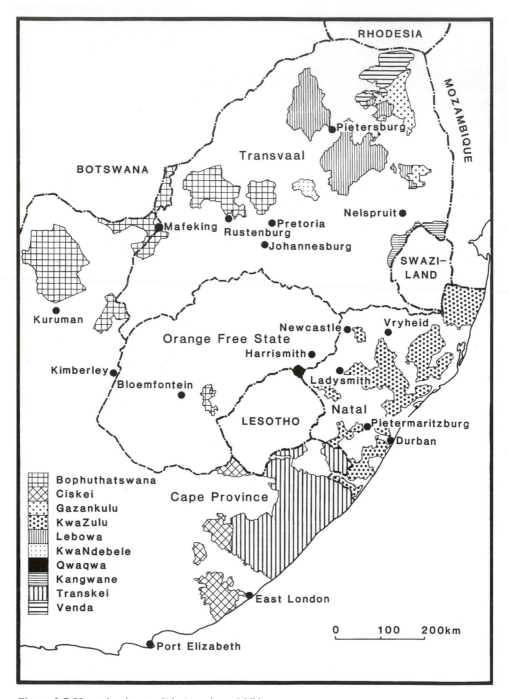

RHODESIA

MOZAMBIQUE

BOTSWANA

Transvaal

●Pietersburg

Nelspruit ●

●Pretoria
Mafeking● Rustenburg
●Johannesburg

SWAZI-
LAND

Kuruman ●

Newcastle ●   ● Vryheid
Orange Free State
Harrismith ●
Kimberley ●   ● Ladysmith
●Bloemfontein   Natal

LESOTHO   Pietermaritzburg
                 ●Durban

Bophuthatswana
Ciskei
Gazankulu   Cape Province
KwaZulu
Lebowa
KwaNdebele
Qwaqwa
Kangwane
Transkei
Venda

                        0   100   200km

                        ● East London

●Port Elizabeth

Figure 3.7 Homeland consolidation plans, 1973
Source: After A. Lemon (1976) *Apartheid: A Geography of Separation*, Farnborough: Saxon House

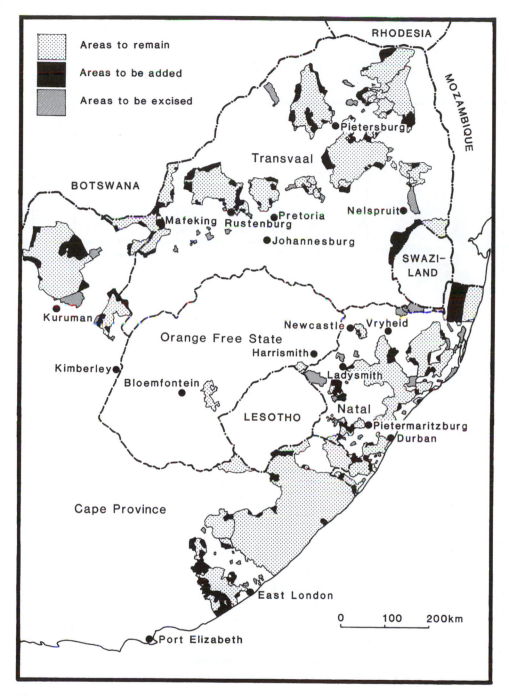

Figure 3.8 Homeland consolidation plans, 1975
Source: After A. Lemon (1976) *Apartheid: A Geography of Separation*, Farnborough: Saxon House

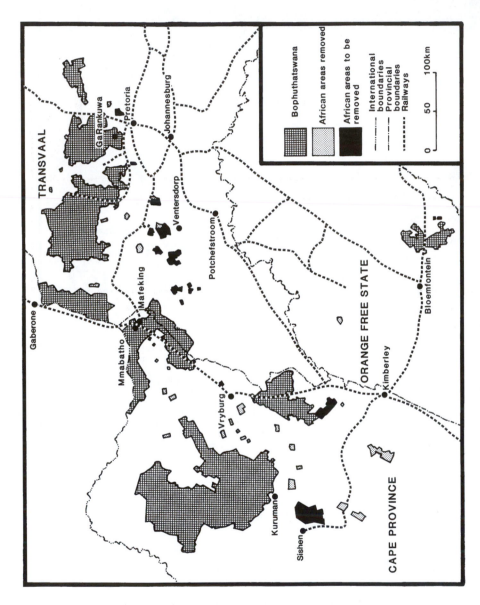

Figure 3.9 Consolidation of Bophuthatswana, 1972
Source: After H.J. Moolman (1977) 'The creation of living space and homeland consolidation with reference to Bophuthatswana', *South African Journal of African Affairs* 7: 149–66

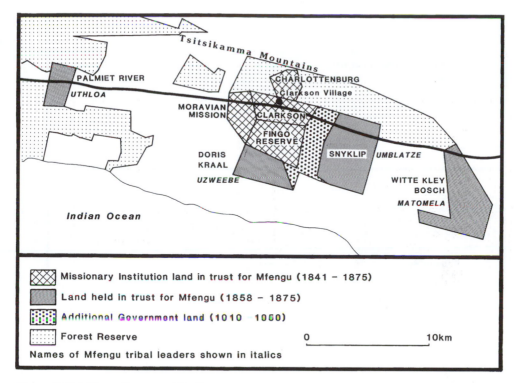

Figure 3.10 Mfengu Reserves, Tsitsikamma
Source: Compiled from information extracted from the Deeds Office and Surveyor-General's Office, Cape Town

Organization of African Unity, especially as it was being undertaken to foster apartheid. The scheme was abandoned with the death of King Sobhuza II of Swaziland later in the same year.

RESETTLEMENT

The government also embarked upon a programme of resettling African people from the designated White areas of the country in the homelands. The major movement was the resettlement of those considered to be surplus to the needs of the White farming community. The provisions of the 1913 Natives Land Act and subsequent measures were directed towards reducing African occupation levels on White-owned lands. The practices of labour tenancy and share-cropping were also abolished. This led to a substantial reduction in the number of African families living on White-owned farms, as both practices involved the presence of Africans on White-owned land as semi-independent farmers rather than as wage labourers. The continued presence of independent African farmers on White-owned farms not occupied by their owners was also condemned and regulations enforced to end such occupation. A commission of inquiry was able to produce a number of maps which evoked marked political emotions (Figure 3.12) (South Africa 1960).

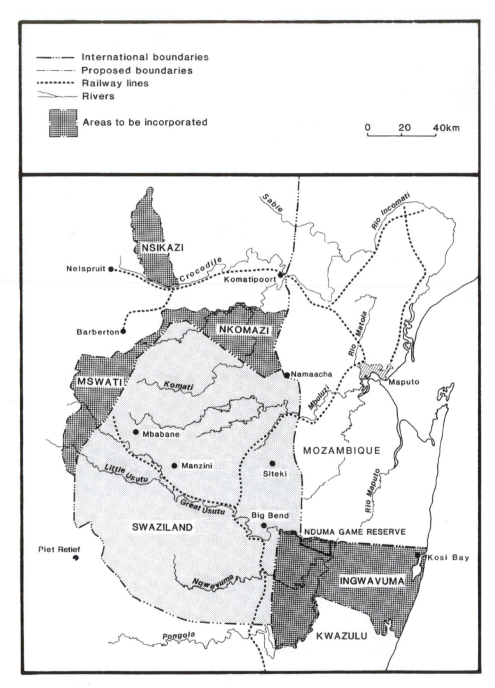

Figure 3.11 Swaziland extension proposals, 1982
Source: After P. Esterhuysen (1982) 'Greater Swaziland?', *Africa Insight* 12: 181–8

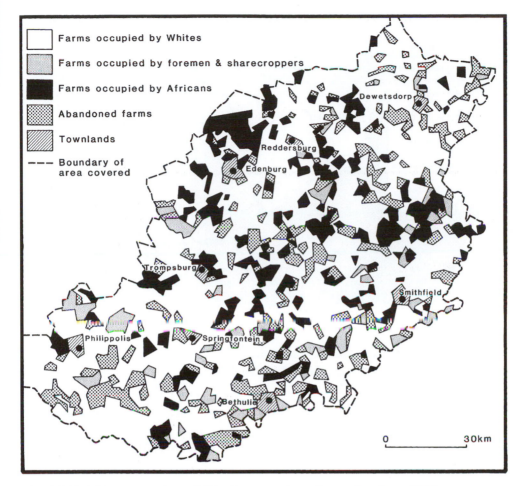

Figure 3.12 African occupation of White farms, southern Orange Free State, 1959
Source: After South Africa (1960) *Report of the Commission of Inquiry into the European Occupancy of the Rural Areas*, Pretoria: Government Printer

The government accordingly sought two goals in the removal of Africans to the homelands. The first was the aim of 'whitening' the rural areas of White South Africa. In 1921 Africans, Coloureds and Asians outnumbered Whites in the White-owned rural areas by a ratio of 3:1. As a result of White rural depopulation, and the continued growth of the African and Coloured rural population, by 1960 the ratio approached 9:1 (Figure 3.13). White political complaints about the 'blackening' of the White rural areas was a major spur to official action. The second goal was the converse increase in the number of Africans living in their own designated national states and hence subjects of the homeland governments. Thus the proportion of Africans living within the African states had risen to over 60 per cent by 1985.

Estimates produced by the Surplus People Project in 1985 suggested that over 1.7 million people were displaced under this programme between 1960 and 1983 alone (Figure 3.14) (Platzky and Walker 1985; Unterhalter 1987). The removals varied from family

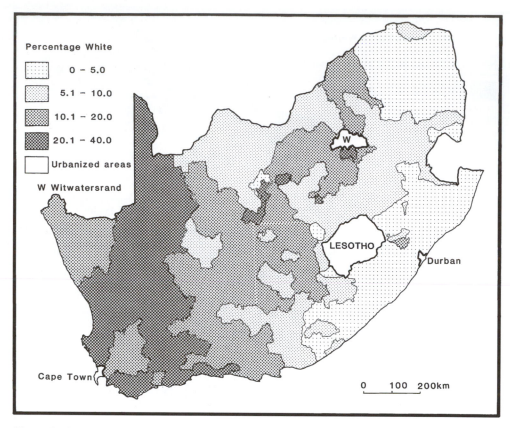

Figure 3.13 Percentage of rural population classified White, 1960
Source: Based on statistics in South Africa (1963) *Population Census 1960,* Pretoria: Government Printer

displacements by individual White farmers to large-scale forced removals organized and executed by the government, often with police and military coercion (Murray and O'Regan 1990). Substantial movements took place particularly from the central Cape, which was perceived as a Coloured labour preference region (see Figure 4.17, p. 122). The numbers were swollen by those displaced from the urban areas under the various influx control measures, who were also forced to move to the homelands.

Resettlement programmes in the African states were of varying quality, often reflecting the nature of the displacement from the White farming areas. Frequently, no prior preparation of the reception sites was made and tents or other improvised shelters awaited those displaced. In other cases, resettlement villages were laid out and prepared with basic services before the arrival of the deportees of some of the larger organized programmes. Usually only sites were laid out before arrival. The result was a series of major resettlement camps such as Dimbaza in Ciskei, where Father Cosmas Desmond first drew the attention of the international community to the poor conditions in the camps and the plight of the refugees (Desmond 1970).

Resettlement was not a popular policy even for the African homeland governments. Supplying the needs of the new arrivals placed considerable strains upon the limited resources of

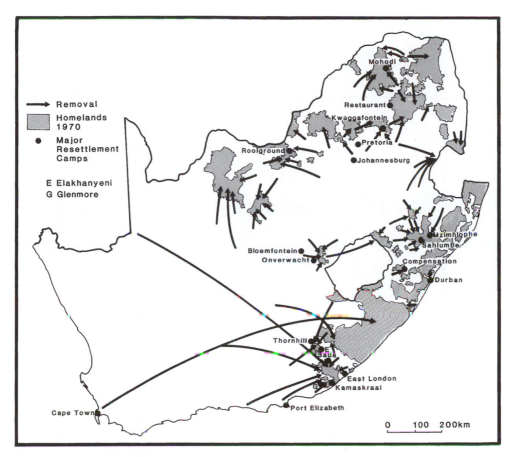

Figure 3.14 Forced removals
Source: Compiled from information in the regional reports of the Surplus People Project

the administrations. Hence the South African government often chose to locate resettlement camps on Development Trust land adjacent to the homelands and then negotiate their inclusion. Accordingly, Ciskei was the scene of most of the Xhosa resettlements from the Cape Province as the Transkei government refused to co-operate and little Trust Land was incorporated into the state. In contrast the KwaNdebele homeland consisted almost entirely of resettlement camps organized by the South African government and then transferred to the new state (Figure 3.15).

Furthermore, the new arrivals often had no immediate place within the power structures which had been evolved in the early stages of the homeland states' existence. Thus the refugees often remained among the poorest communities in the states and were generally without access to land either for grazing or cultivation. Also few refugees were placed on the various agricultural schemes undertaken in the homelands, which were reserved for those considered to be long-term and politically reliable citizens. Even those enticed to move from one homeland to another found the enticements impossible to gain in reality. The approximately 50,000 Herschel and Glen Grey refugees from Transkei to Ciskei in the post-1976 era fared little better than others displaced from White South Africa.

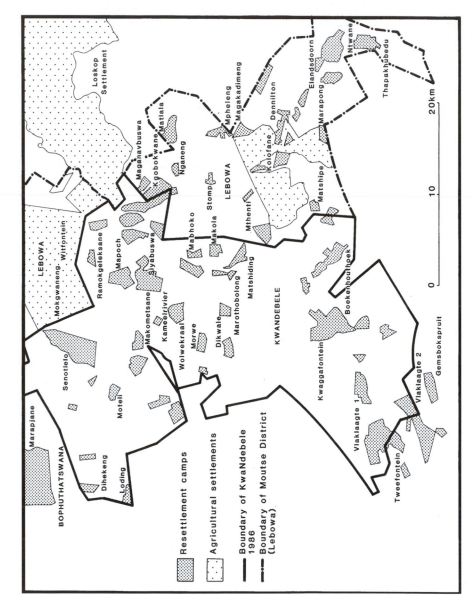

Figure 3.15 Resettlement camps in KwaNdebele
Source: Compiled from official 1:250,000 and 1:50,000 topo-cadastral maps

HOMELAND STRUCTURES

The homeland governments established their own administrations, which took over the extensive powers previously exercised by the Native Affairs Commissioners and the Department of Native Affairs and Education. Executive authorities headed by Chief Ministers of self-governing states or Presidents of independent states were served by a range of ministries, which gradually assumed the running of the administration (Butler *et al.* 1977). Bureaucracies provided significant areas of expenditure and employment for the African population. Consequently, one of the immediate problems faced by all the homeland governments was the establishment of an administrative headquarters. As few towns were included within the homeland boundaries this constituted a major problem. Provisionally, the offices of the local Commissioner in the main nearby White town were used. Then new offices and governmental complexes were built in the African township attached to the major White town in the vicinity. In this manner Zwelitsha outside King William's Town and Seshego outside Pietersburg became the provisional capitals of Ciskei and Lebowa respectively.

The South African government drew up a set of guidelines for the identification of new permanent sites for the homeland administrations (Best and Young 1972). They called basically for the designation of an open waste area, removed from White influence, without historic significance, which would be central to the future state area and population distribution. Further, it was suggested that the site lie on a development axis in terms of national development programmes. The nine homeland governments, joined by KwaNdebele in 1978, were then offered the official choices (Figure 3.16). It is worth noting that the requirement that the sites be without historic significance was rejected by several homeland governments once they gained sufficient power to influence state policies. Thus the South African government's choice of Nongoma for the KwaZulu capital was rejected in favour of historic Ulundi, site of the last royal capital of independent Zululand before 1879.

The central government's choice of Debe Nek as capital of Ciskei was rejected by the Ciskei government in favour of Alice, the seat of the University of Fort Hare and Lovedale College, the educational hearth of the Xhosa people. In other states only one suitable site was available. The restricted territorial extent of the Qwaqwa homeland was such that Witsieshoek, duly renamed Phuthaditjhaba, was the only site physically available. Most significantly, Umtata was designated the capital of Transkei, although nominally a White town, but entirely surrounded by African rural reserves. This was to cause considerable controversy as other homeland governments claimed the major town associated with their state. Consequently, the Ciskeian demand for the incorporation of King William's Town was rejected by the South African government, although the Bophuthatswana government's demand for the incorporation of Mafeking was acceded to (Parnell 1986). As a compromise the Ciskeian government finally selected Bisho, adjacent to King William's Town, as its permanent site. It had little in common with any of the guidelines laid down by the South African government, being within 3 kilometres of a White town, eccentric to the national territory and population, and within a restricted site. It did serve notice, however, that King William's Town would be regarded as future Ciskeian territory (Figure 3.17).

The new state capitals were the subject of extensive conspicuous spending as the various governments established new complexes, official housing and the infrastructure commensurate with a state capital. In few cases was there an inherited infrastructure. New sites offered considerable scope for grandiose town planning schemes. Thus Lebowakgomo, Mmabatho and Bisho were laid out with extensive government quarters, including a presidential palace, parliament buildings and government ministries. Expansive housing schemes

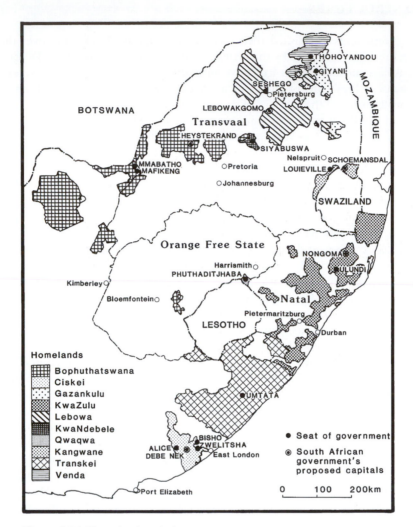

Figure 3.16 Homeland capitals
Source: Modified after A.C.G. Best and B.S. Young (1972) 'Capitals of the homelands',
Journal for Geography 3: 1043–55

supplied a range of housing styles and sizes not previously available for Africans in South
African towns, ranging from the ministerial complex to civil service quarters and the latterly
permitted squatter areas (Figure 3.18). Commercial and industrial sectors were also designated, in order to generate employment and promote the capital as the major town within
the various homelands' urban hierarchies.

DECENTRALIZATION

Economic development was essential to support a greater proportion of the African
population within the homelands. The Tomlinson Commission had stressed the need for

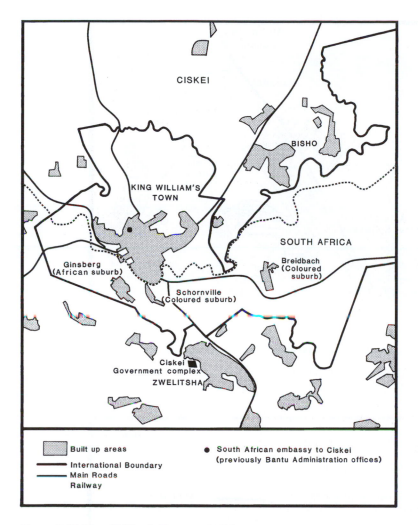

Figure 3.17 King William's Town and Bisho
Source: Compiled from official 1:250,000 and 1:50,000 topo-cadastral maps

substantial investment in the homelands to prevent a massive flow of population to the (White) cities. However, the government showed little inclination to spend the sums required and only gradually developed a cohesive policy. A major philosophical problem had to be faced in the formulation of a homeland industrial policy. The possibility of White-owned capital being used to invest in industries in the homelands was initially rejected as it was considered that only African-owned capital should be used in order to preserve African independence. Little such capital was available in the 1950s and 1960s, and hence few jobs could be generated.

At the same time the legal barriers to the migration of Africans to the metropolitan areas were tightened and restrictions on the number of African job opportunities were imposed. Strict limits were placed upon the establishment or extension of industrial premises in the metropolitan regions (Figure 3.19). By 1971 the restrictions had been tightened so that no

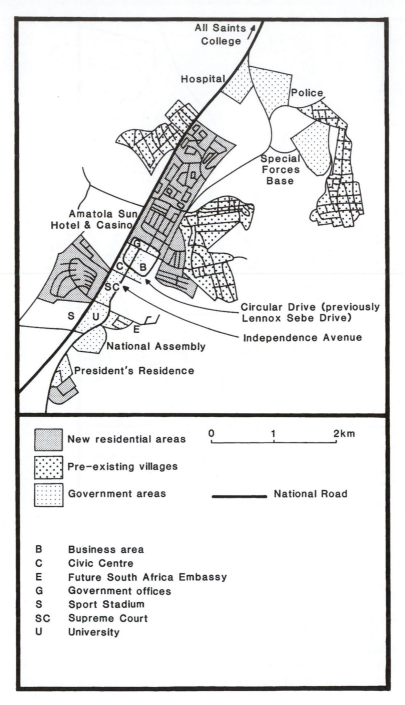

Figure 3.18 Bisho
Source: Based on maps supplied by Bisho Municipality

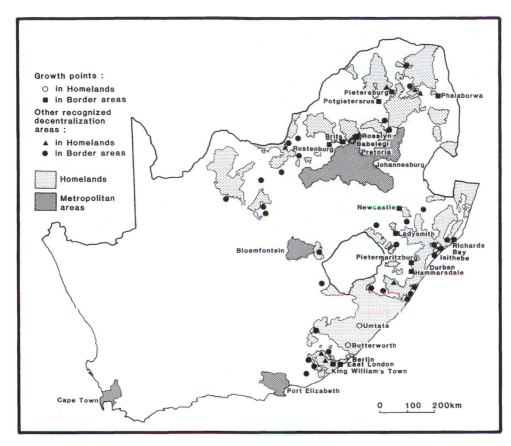

Figure 3.19 Decentralization policies, 1960–73
Source: Based on information in T. Malan and P.S. Hattingh (1976) *Black Homelands in South Africa*, Pretoria: Africa Institute

new industry in the Pretoria-Witwatersrand-Vaal Triangle (PWV) region could employ more than two Africans for every White employee.

Parallel with this policy was the border industry programme begun in 1960 under the auspices of the Permanent Committee for the Location of Industry, later (1971) renamed the Decentralization Board. The object was to develop industrial areas in White South Africa, adjacent to the homelands, so that African workers could commute on a daily basis between their homes and places of work (Bell 1973). The programme overcame the ideological problem of White-owned capital entering African areas, and also meant that African workers could live within the homelands.

Initially the Committee concentrated on those areas likely to attract industrialists in order to demonstrate the practicability of the policy, through the offer of financial and other incentives to establish factories in border cities. Three sites were developed, Hammarsdale, between Durban and Pietermaritzburg, Rosslyn, near Pretoria, and Pietermaritzburg itself. The Durban metropolitan area was excluded as incentives were not required to attract industrialists to that region.

The initial phase was regarded as a success by the government, and in 1968 the geographical spread of sites was widened to include towns adjacent to other homelands. The range of incentives was also extended. The rate of new job creation, however, declined, suggesting that the more attractive sites possibly did not need to be subsidized. In the course of the 1960s some 87,000 jobs were created in border areas, far short of the 20,000 per annum thought to be essential by the Tomlinson Commission in the early 1950s.

It should be noted that the number of cross-border commuters increased as the African townships built in the homelands expanded when limits were placed on the growth of those within White South Africa. Accordingly, Pretoria received an increasing number of African workers from neighbouring Bophuthatswana, and subsequently also from KwaNdebele (Figure 3.20) (Bernstein and McCarthy 1990). Long-distance commuting by bus and taxi on a daily basis became a significant aspect of African urban life as an increasing number of workers were forced to live in the homelands, and commuting times were measured in hours rather than minutes.

HOMELAND GROWTH POINTS

In 1968 the government passed the Promotion of Economic Development of the Homelands Act, which provided for the controlled introduction of White-owned capital into the homelands. Under this Act it was possible for Whites to work as agents or contractors for the South African Bantu Trust or the various homeland development corporations, providing the industrial or mining development became the property of the Trust or development corporation after a stipulated period of time (usually twenty-five years). With the philosophical objections to White capital entering African areas overcome, the identification of growth points for industrial decentralization began. The industrial decentralization programme underwent a number of changes in the course of the 1970s and 1980s, but the purpose remained the same. The scatter of development points was wide and inevitably only a minority were successful in attracting more than a few industries (Malan and Hattingh 1976). The major growth points were in Bophuthatswana, notably at Babelegi, and in KwaZulu, notably at Isithebe (see Figure 3.19). Again, because of their proximity to the industrial heartlands of the Pretoria-Witwatersrand-Vaal Triangle and Durban, this did little to decentralize South African industry. Other more remote centres such as Butterworth in Transkei and Dimbaza in Ciskei were developed on the basis of generous government incentives. In a number of cases sites were only indicated vaguely on a map in order to provide each of the homelands with at least one decentralization point for political rather than economic reasons.

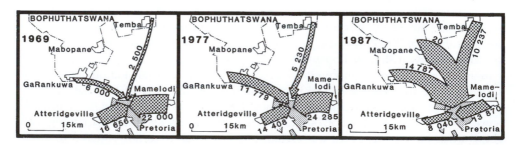

Figure 3.20 Changing patterns of daily African commuter traffic in Pretoria, 1969–87
Source: After A. Bernstein and J. McCarthy (1990) *Opening the Cities*, Durban: Indicator Project

The incentives took the form of foundation grants, labour, power and transport subsidies and tax holidays, and so contributed to some measure of employment. Long-term economic viability was rarely a consideration for many industrial concerns. The policy led to several unexpected results. In the southern Transvaal large numbers of workers were attracted to the development points, but they were often not citizens of the homeland concerned. Thus Babelegi, though in Bophuthatswana, only housed a Tswana minority. Tensions arose in subsequent attempts to impose such rules as the use of Tswana as a medium of primary school education upon the non-Tswana population. The industrial development corporations also sought to attract overseas investment, particularly from South-East Asia. Marked concentrations of Taiwanese and Hong Kong investments became evident. The capital remained essentially footloose, attracted by cheap labour and subsidies, and many factories were closed when the subsidies were withdrawn, particularly in the post-1991 era.

In 1975 the National Physical Development Plan sought to spread economic growth yet more widely with the introduction of the concept of development axes (Figure 3.21). New growth poles were identified to attract industries, and hence workers, away from the southern Transvaal metropolitan region (Fair 1982). In 1982 the decentralization programmes were brought within a national framework of the revised National Development Plan. The multiplicity of development points remained, but attention was directed towards a few potentially more successful deconcentration points (Figure 3.22). Although a number of the deconcentration points were situated in the western parts of the country, as indeed had been some decentralization points, attention was primarily directed towards the homelands. Thus Ekangala, adjacent to the Ekindustria industrial area, was declared a deconcentration point in an effort to promote the development of the KwaNdebele homeland. Similarly Botshabelo was designated as a non-Tswana deconcentration point primarily for the South Sotho population of Bloemfontein and those displaced from farms in the eastern Orange Free State. The designation of fewer points with greater incentives was expected to offer advantages, notably in an effort to boost economic growth rates in the economic heartland, which remained the driving force of the South African economy. The slowdown of the South African economy in the late apartheid period thus necessitated a review of the dispersed decentralization effort and its concentration in fewer but more viable centres.

HOMELAND POVERTY

One of the most prominent aspects of the homelands was their poverty, a serious heritage bequeathed to the post-apartheid government. Resettlement of unemployed people, high natural population growth rates, agricultural stagnation and limited industrial growth all contributed to the perpetuation of a cycle of poverty. Without anywhere to go and without access to international capital on any scale, the majority of the homeland populations exhibited remarkably low incomes, although there were very limited areas of prosperity.

Estimates produced by the Development Bank of Southern Africa for 1985 indicated very substantial disparities in incomes (Leistner 1987). Whereas the Gross Domestic Product per head in the homelands varied from approximately R600 to R150, that of the remainder of South Africa was approximately R7,500 per head (Figure 3.23). Only the development of mining in Bophuthatswana provided an income by the 1980s which indicated any degree of prosperity. Bophuthatswana also benefited from its proximity to the industrial heartland

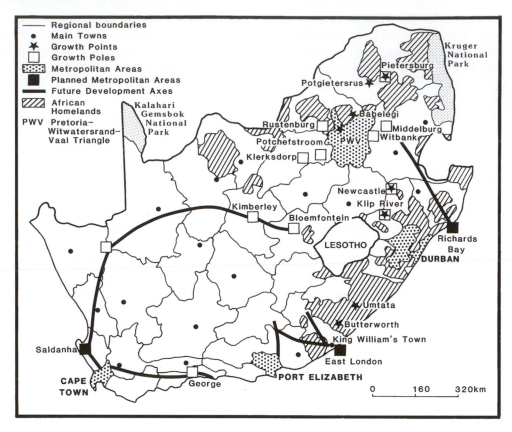

Figure 3.21 National Physical Development Plan, 1975
Source: After T.J.D. Fair (1982) *South Africa: Spatial Frameworks for Development*, Cape Town: Juta

of the country as the site of the major decentralization and deconcentration points. The disparities were narrowed through government subventions to the homeland governments and the remittances of migrant workers, which accounted for up to three-quarters of Gross National Product. The relative position of the African states deteriorated as a result of the resettlement schemes and the imposition of influx control, which prevented a poverty-stricken rural population from migrating to the cities.

Nevertheless, one of the features of poverty was the high rates of temporary short–term labour migrancy from the homelands to White South Africa, which various official pro-grammes of decentralization had attempted to reduce. The impact of the migrant labour system affected the entire subcontinent (see Figure 7.13, p. 192). However, within South Africa the effects upon the various homelands differed substantially (Figure 3.24). In 1970 only Gazankulu recorded a majority of males in the vital 15–64 age group temporarily absent from their homes. Generally those homelands furthest from the industrial areas recorded a higher ratio of absentees to residents. Thus Transkei, with virtually no opportun-ities for cross-border commuter traffic, recorded a substantially higher ratio than Ciskei, where the East London industrial complex and other towns enabled a larger proportion of workers to become commuters rather than migrants.

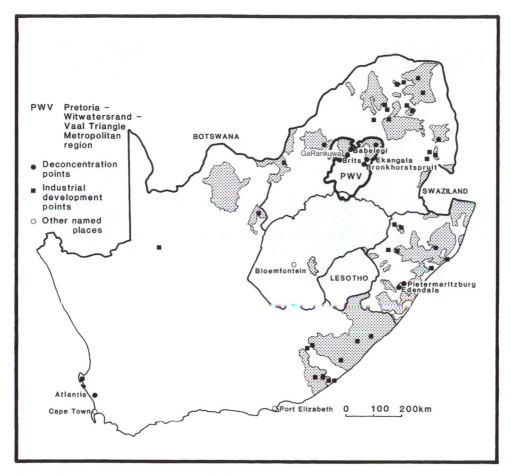

Figure 3.22 Deconcentration points, 1985
Source: Based on map published by the Development Bank of Southern Africa

CASINO STATES

The independence of four of the homelands released them from a host of South African legal restrictions, including those related to gambling. The National Party government retained a highly restrictive attitude towards gambling and other activities it regarded as morally dubious. The independent homelands provided the venue for allowing wealthy South Africans to indulge themselves, free of the restrictions imposed by the Separate Amenities Act, the Immorality Act and the gambling laws, without having to leave either the country or the Rand currency area. In this they followed the example set by Botswana, Lesotho and Swaziland in the late 1960s, which had encouraged casino development in order to boost the local tourist industry.

A network of casino resorts was designed initially by the Southern Sun Hotel chain of South Africa (Figure 3.25). The result was the construction of a spread of enterprises strategically placed with regard to the main centres of the country (Stern 1987). The

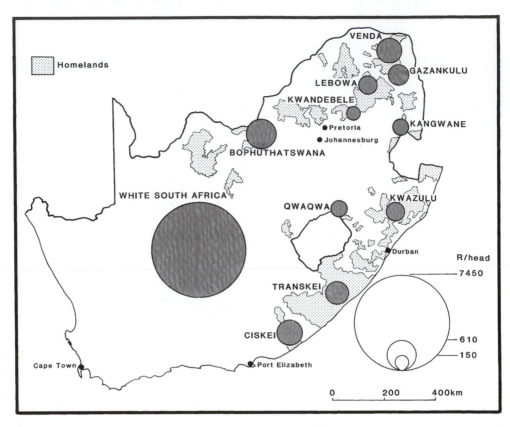

Figure 3.23 Gross Domestic Product per head by state, 1985
Source: Based on statistics of the Development Bank of Southern Africa (1985) in G.M.E.
Leistner (1987) 'Africa at a glance', *Africa Insight* 17: 1–104

major resort was developed at Sun City in Bophuthatswana, only 120 kilometres from Johannesburg and Pretoria. Because of its situation, it was expanded several times, culminating in the creation of the exotic Palace of the Lost City. The adjacent country club, with its leisure facilities and world-renowned golf course, formed part of a comprehensive and luxurious holiday resort. Other casino resorts were established in Bophuthatswana which catered for a wide geographical spread of trade including GaRankuwa, north of Pretoria, and Thaba Nchu near Bloemfontein. The first casino in the Transkei was eccentrically situated on the Wild Coast adjacent to the Natal South Coast, one of the major holiday regions of South Africa. Ciskei similarly developed not only the urban casino in Bisho but a casino resort on the Fish River boundary to attract people from the eastern Cape.

POSTSCRIPT TO STATE PARTITION

State partition has been a most controversial policy whenever it has been pursued in the twentieth century, whether in India, Cyprus or Palestine (Christopher 1999). The ethnic basis of state partition plans has always been regarded as suspect by those groups which lost

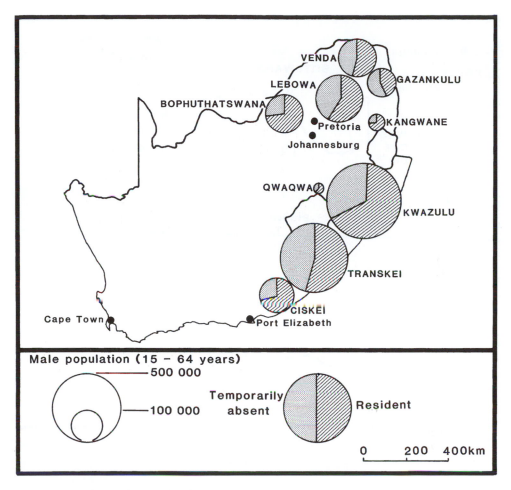

Figure 3.24 Labour migrancy levels, 1970
Source: Based on statistics in T. Malan and P.S. Hattingh (1976) *Black Homelands in South Africa,*
Pretoria: Africa Institute

political powers or lands. There has been much argument concerning the basis of the South
African National Party government's plans to partition the country. The basic premise that
partition was necessary for national survival was held not by the majority of the population,
only by a section, if a dominant section, of the ruling White minority for part of the twentieth century. Theoretical academic and political alternatives to the government's partition
plans were suggested periodically to provide a more equitable distribution of the country's
land resources. They are worthy of examination if only to expose the essential impracticality
of the entire programme.

The arguments regarding a more equitable partition were largely academic and dismissed
in political circles as being too idealistic and expensive. Thus Blenck and von der Ropp
(1977) proposed a scheme for the division of the country into African and non-African
halves (Figure 3.26). The basic argument was the acceptance that the African and White
views of the future nature of the South African state were incompatible and that a radical

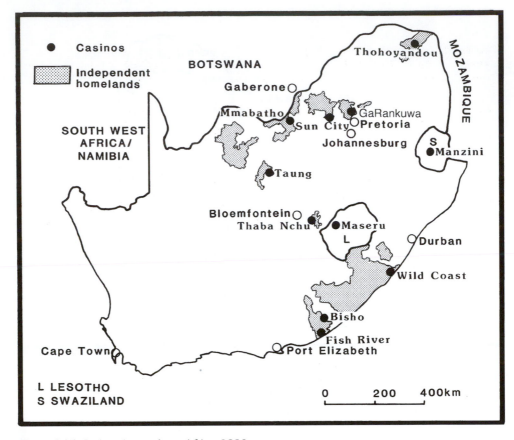

Figure 3.25 Casinos in southern Africa, 1990
Source: Based on information supplied by Sun Hotels

partition acceptable to both sides was worth investigation. The African–non-African division appeared as a variant of the indigenous–immigrant dichotomy, assuming that the Coloured and Indian population would identify with the Whites if offered full political and economic rights. The boundary reflected the historic frontier delimiting the extent of African settlement, which coincided with regions of low economic development. Maasdorp (1980) later refined the idea by indicating the intricacy of the border zone and suggesting a compromise line. Within the two states accordingly identified, Capeland and Capricornia, the White population would have been in a minority to a Coloured majority in the former and to an African majority in the latter. The basic element of these schemes was the fact that the Witwatersrand and Natal would fall within the African state, clearly a viable practical policy only for a few White political parties.

By the late 1980s the anticipated demise of the apartheid era again raised the policy of partition as a political platform (du Toit 1991). The scale of the rump White state varied from the Conservative Party's attempt to maintain the homeland system with only a few modifications, to the Oranjewerkers who initially sought a single village (Morgenzon, in the eastern Transvaal) where they could pursue an all-White labour programme, and so preclude all contact with the surrounding African population except at official or commercial levels.

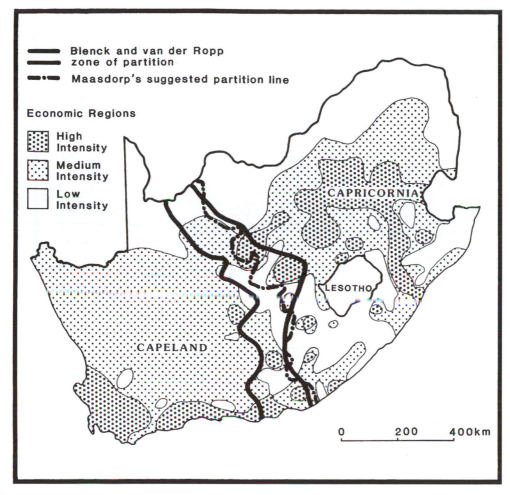

Figure 3.26 Theoretical partition plans
Source: Modified after J. Blenck and K. von der Ropp (1977) 'Republic of South Africa: is partition a solution?', *South African Journal of African Affairs* 7: 21–32, and G. Maasdorp (1980) 'Forms of partition', pp. 107–46 in R.I. Rotberg and J. Barratt (eds) *Conflict and Compromise in South Africa*, Cape Town: David Philip

Three widely publicized schemes sought to create not a White state but an Afrikaner state. First the Afrikaner Resistance Movement (AWB) propounded the concept of a Boerestaat (Farmers' State) for the Afrikaner nation. This was narrowly focused on territories of the two former republics of the Transvaal (South African Republic) and the Orange Free State, which had fought against Great Britain between 1899 and 1902 (see Figure 1.6, p. 16). The area of northern Natal transferred from the Transvaal to Natal at the end of the war remained a territorial claim for such a state, although the interests of the Zulu nation were recognized and respected. A variant of this scheme was the more narrowly drawn state excluding the African homelands and areas of the country such as the central Witwatersrand and Natal which were considered to be heavily Anglicized. The result was the proposal of the Afrikanerland of the Oranjewerkers movement (Figure 3.27). The proposed state retained

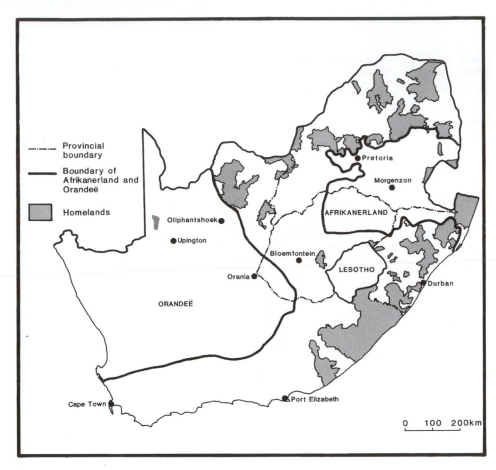

Figure 3.27 Right-wing White political partition plans
Source: Modified after maps in *Sunday Times*, 19 February 1989, Johannesburg, and *Time*, 17 December 1990, New York

much of the agricultural heartland of the country as well as much of the industry, and a coastline.

In contrast the state of Orandeë proposed by the Afrikaner Freedom Foundation did not overlap Afrikanerland at all. It was proposed to provide an isolated desert state, which through its very inhospitableness would not be regarded as either desirable or a threat to the African government which it was assumed would come to power in the remainder of the country. The expounded philosophical basis of the state paralleled that of Israel, in the belief that all Afrikaners would feel more secure if they possessed a homeland in which their culture and language would be preserved by law. It was not intended that more than a small proportion of the Afrikaner nation would reside in Orandeë. The village of Orania on the Orange River was purchased by the Foundation and attempts were made to put the Afrikaner national development programme into practice.

4 Urban apartheid

Apartheid operated on a local, urban level. The major impetus of the initial years of the National Party administration after 1948 was directed towards the implementation of residential and personal segregation rather than the more philosophically based concept of state partition. Much of the basic legislation affecting Africans was already on the statute book and was duly enforced with greater ruthlessness. However, the residential segregation of the remainder of the population was a popular issue which enjoyed wide support within the White community of whatever political persuasion. By the 1960s political debate in the country, whether in White or African communities, was essentially urban based. Urban segregation and the attempts to restrict African access to the urban areas were thus the dominant political issues of the ensuing decades.

THE POPULATION REGISTRATION ACT

In 1950 the two major pieces of legislation designed to create the new apartheid city were enacted, namely the Population Registration Act and the Group Areas Act. The Population Registration Act provided for the compulsory classification of the population into discrete racially defined groups. Three basic groups were identified: White (European), African (Bantu or Black) and Coloured. The latter group was split into several subdivisions: Cape Malay, Griqua, Indian, Chinese and a residual Cape Coloured group. The *ad hoc* and often ambiguous classifications adopted in the colonial and early Union period thus gave way to a rigid system, enforced by statute and specified on identity documents.

The criteria adopted for classification purposes were based on physical appearance and social acceptability. The Act stated that:

> 'White person' means a person who in appearance obviously is, or who is generally accepted as a white person, but does not include a person, who, although in appearance obviously a white person, is generally accepted as a coloured person.
>
> (South Africa 1950: 277)

Thus in the process of classification considerable latitude was theoretically possible, although, in general, the rules were applied to prevent anyone who was not recognizably *and* socially acceptable as a White person from gaining that status. Families were consequently split as members were subjected to such distasteful tests as curliness of hair, skin colour and linguistic ability. The Act also made provision for reclassification where it was considered that descriptions were inaccurate. By the late 1980s as many as 1,000 people a year were seeking reclassification – mostly to become Coloured instead of African or White instead of Coloured. Indeed, between 1983 and 1990 some 7,000 persons had their racial description changed (Figure 4.1). In this period only 120 people changed their classification to African, but 3,561 changed to White. It might also be noted that the children of mixed marriages nearly always took the race of the parent down the political pecking order, thereby keeping the White group as White as possible, although if there was no White parent the classification of the father was often adopted. Under the Prohibition of Mixed Marriages Act (1949) marriages between Whites and members of other groups were prohibited. In 1950 the Immorality Amendment Act banned extra-marital relations between Whites and members of all other groups. (Relations between Whites and Africans had been made illegal in 1927.) The various prohibitions did not affect marriages and extra-marital relations between members of other groups, as only the racial purity of the White group was of official concern.

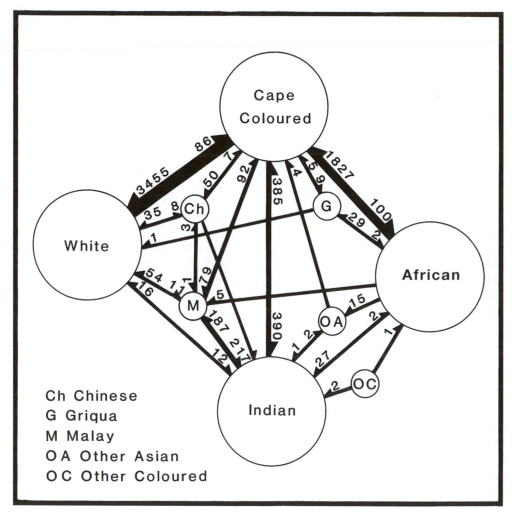

Figure 4.1 Changes in race classification, 1983–90
Source: Based on statistics in *Hansard*, 19 June 1991, col. Q2009

THE GROUP AREAS ACT

The Group Areas Act was referred to by the Minister of the Interior, Dr T.E. Donges, as 'one of the major measures designed to preserve White South Africa', when he introduced the Bill in parliament (*Hansard* 1950: col. 7722). Its conception was to effect the total urban spatial segregation of the various population groups defined under the Population Registration Act. Towns and cities were to be divided into group areas for the exclusive ownership and occupation of a designated group. People not of the prescribed designated group would be forced to leave and take up residence in the group area set aside for their own group. The result was to be total segregation (apartheid), not the piecemeal results of colonial and Union segregationism. Thus social contact between the communities would be reduced to a minimum and competition for urban space legally eliminated.

Administratively, the Land Tenure Advisory Board, established in 1946 to oversee the drawing of Indian areas, assumed the task of drawing up the comprehensive plans for the new apartheid cities (Mabin 1992). In 1955 it was renamed the Group Areas Board. The process of establishing group areas was long and complex, involving the formulation of the municipalities' proposals, public inquiries, Board recommendations and ministerial approval before the final proclamation was published in the *Government Gazette*. Furthermore, the initiation of schemes was prompted by the central authorities, often in the face of municipal opposition. However, the reordering of towns and cities was part of the general philosophy of town planning in the post-Second World War era, and sustained opposition was rare. Moreover, the administrative machinery for effecting the purposes of the Act was not put in place until 1955, when the Group Areas Development Act (after 1966 the Community Development Act) provided a mechanism for expropriation and land development.

Although the supporters of the Act clearly envisaged the inclusion of the African population within its ambit, the Ministry of Native Affairs and its successors were sufficiently powerful to retain control over the African population through separate legislation. The Natives Resettlement Act of 1954 provided the mechanisms required to remove Africans from inner African freehold areas, while the existing legislation and its numerous amendments enabled the government to control virtually all aspects of African personal and community life. The practical problems of declaring African group areas in an era when Africans could not own land in freehold tenure outside the homelands further confused the issue. The attempt to create a uniform system after the 1950s only resurfaced briefly in the late 1980s and was overtaken by events. The Group Areas Act for most purposes therefore was only specifically applied to the non-African population, although the exclusionary clauses did affect Africans. This most notably applied to the periods of time given to people disqualified under the Act to evacuate or sell properties in proclaimed group areas.

The guidelines for demarcating group area boundaries were drawn up by the Durban Corporation in the light of its experience in segregating the Indian population before 1950, and were informally adopted by the Land Tenure Advisory Board (Western 1996). The guidelines proposed that group areas be drawn on a sectoral pattern with compact blocks of land for each group, capable of extension outwards as the city grew. Group areas were to be separated by buffer strips of open land at least 30 metres wide, which were to act as barriers to movement and therefore restrict social contact. Accordingly rivers, ridges, industrial areas, railways, etc. were incorporated into the town plan. Links between the different group areas were to be limited, preferably with no direct roads between the different group areas, but access only to commonly used parts of the city, for example the industrial or central business districts. Originally it was also intended that each group area should be self-sufficient in shopping facilities and that local self-government be introduced. The guidelines were subsequently systematized to form the model apartheid city (Figure 4.2) (Davies 1981).

In practical terms, given the White dominance in the Board's deliberations, a number of salient points emerge. First, the city centres, with virtually no exceptions, were zoned as part of the White group area. Second, areas zoned for the other groups were highly restricted and peripheral. Third, the sectoral zonings affected not only racially mixed residential areas, but many previously segregated suburbs which fell within the broad divisions of another group.

Armed with these guidelines and a host of plans, proposals and suggestions from municipalities, political parties and other interested parties, the Land Tenure Advisory Board began the redrawing of South African cities (Pirie 1984). If the case of Port Elizabeth is examined, there was a gap of ten years between the drawing up of the first group area proposals (1951) and the first proclamation (1961) (Davies 1971). In that time the local authorities consulted widely, and offered varying schemes which failed to meet the Land Tenure Advisory Board's requirements of neatness and sectoral contiguity. Those of the Joint Town Planning Committee clearly deviated from the model quite widely (Figure 4.3). The retention of isolated pockets of Coloured or Asian housing which the Municipality wished to leave undisturbed was equally not acceptable to the Board (Figure 4.4).

Ultimately the Board imposed its own proposals upon the Municipality (Figure 4.5). The final proclamation, therefore, not only adhered closely to the guidelines but specified stringent evacuation periods for those disqualified, ranging from one to ten years. Additional land was set aside as future group areas. Significantly, the separate Malay area which had been the subject of considerable ingenuity in preliminary planning was abandoned, but the separate Chinese area survived and was incorporated into the final plan.

The proclamation plan left Walmer Township isolated from the main block of African suburbs, as a result of the independent status of the Walmer Municipality (See Figure 1.21, p. 34). However, following the latter's administrative incorporation into Port Elizabeth in 1967, the Township was faced with the threat of demolition and the inhabitants with resettlement as part of the broad zoning scheme. A lack of alternative housing constantly delayed the removals and it was only in 1986, after two decades of neglect, that the survival of the Township was accepted by the authorities.

That the 1961 proclamation was not the last word on the subject is attested to by some sixteen subsequent proclamations which extended and modified the original dispensation between 1963 and 1990 (Figure 4.6). The extensive White area was enlarged only slightly but the other group areas required substantial additions as the city grew. Similar problems were experienced in the other major cities, where the pre-apartheid inheritance was complex.

The national rate of proclamation was at first quite slow as few authorities were able to draw up the necessary plans quickly (Christopher 1991). By 1960 only 175,000 hectares had been proclaimed. However, in the course of the 1960s a further 580,000 hectares were designated and the majority of the towns and cities in the country were covered by the proclamations (Figure 4.7). Thereafter zonings generally were related to extensions for new housing areas and modifications arising from changed plans.

The range of interpretations incorporated into the group area plans was substantial, suggesting that the Board's guidelines were only very broadly applicable. In a number of cities the Group Areas Board was intent upon proclaiming as much land as possible for the White population. The proclamation of an extensive area covering Table Mountain in Cape Town as a White group area in 1958 was evidence of the symbolic use of the Act, to exclude other

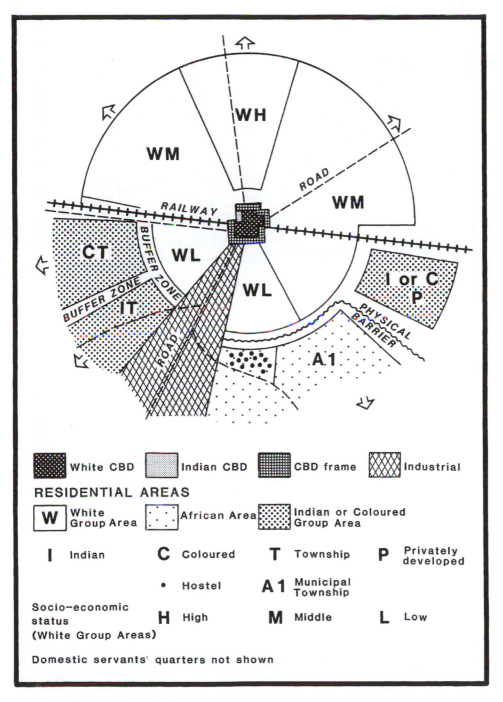

WH

WM

ROAD

WM

RAILWAY

BUFFER ZONE

CT

WL

I or C
P

BUFFER ZONE

BUFFER ZONE

IT

WL

ROAD

PHYSICAL
BARRIER

A1

	White CBD		Indian CBD		CBD frame		Industrial

RESIDENTIAL AREAS

W	White Group Area		African Area		Indian or Coloured Group Area

I	Indian	C	Coloured	T	Township	P	Privately developed
		•	Hostel	A1	Municipal Township		

Socio–economic status (White Group Areas)

H	High	M	Middle	L	Low

Domestic servants' quarters not shown

Figure 4.2 The model apartheid city
Source: After R.J. Davies (1981) 'The spatial formation of the South African city', *GeoJournal* Supplementary Issue 2: 59–72

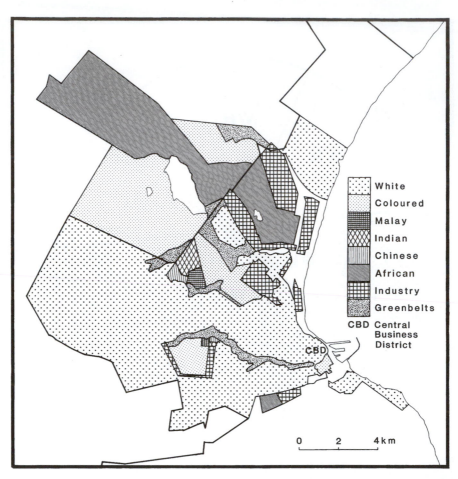

Figure 4.3 Port Elizabeth Joint Planning Committee group area proposals, 1953
Source: After W.J. Davies (1971) *Patterns of Non-White Population Distribution in Port Elizabeth with Special Reference to the Application of the Group Areas Act,* Port Elizabeth: University of Port Elizabeth

people from the use of the area, as no residential development was envisaged. The most extensive White group area (80,000 hectares) was proclaimed around the administrative capital, Pretoria, where the plans for the whole country were drawn up. The wholesale zoning of towns and cities as White group areas was such that in most cases allowance was made for the needs of the White group for the foreseeable future. In other cases proclamations were restricted to existing residential and business areas only. In the mining and industrial towns of the Witwatersrand the result was a patchwork of zoned areas and intervening 'controlled' areas, which effectively remained White by virtue of the White ownership of the land (Figure 4.8).

In contrast, the areas set aside for the other groups were severely prescribed. In general, the Indian and Coloured areas were small and peripheral, with few exceptions. The approach adopted by the Board varied from the division of existing areas of the town into zones predominantly occupied by the various groups, to the proclamation of all the existing areas

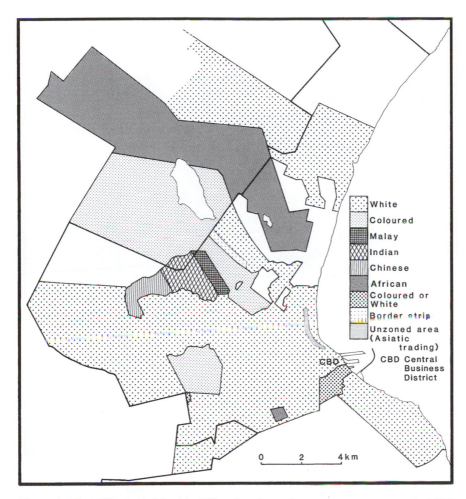

Figure 4.4 Port Elizabeth Municipal Planning Committee group area proposals, 1955
Source: After W.J. Davies (1971) *Patterns of Non-White Population Distribution in Port Elizabeth with Special Reference to the Application of the Group Areas Act*, Port Elizabeth: University of Port Elizabeth

as White and the setting aside of new areas devoid of any buildings for the other groups. In a number of cases group areas were redrawn in order to exclude Indians from the central business districts.

Although provision was made for the establishment of group areas for the Cape Malay, Chinese and Griqua populations, few such areas were ever designated. One Cape Malay area, the Malay Quarter of Cape Town, was established, but the majority of the Cape Malay population, even of that city, had to live in areas designated for the Coloured community. Although several Chinese areas were proclaimed, only one, in Port Elizabeth, was ever developed as the small population was widely scattered in areas such as the Witwatersrand. No Griqua areas were set aside even in East and West Griqualand.

Attempts were made in the Transvaal to centralize the scattered Indian and Coloured urban communities into regional centres. In these centres separate facilities, including

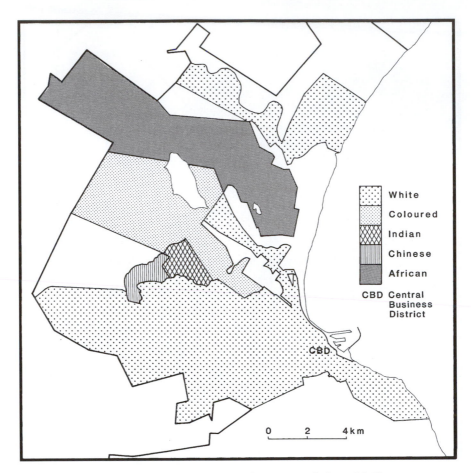

Figure 4.5 Port Elizabeth Group Areas Board recommendations, 1960
Source: After W.J. Davies (1971) *Patterns of Non-White Population Distribution in Port Elizabeth with Special Reference to the Application of the Group Areas Act*, Port Elizabeth: University of Port Elizabeth

schools and clinics, could be provided for communities considered to be too small to support the basic infrastructure of a group area in each small town. Thus many of the initially designated group areas for the two groups established in the late 1950s and early 1960s were abolished in favour of centralization (Figure 4.9). The same form of centralization was attempted on the East and West Rand until the 1980s where Roodepoort and Boksburg established group areas for the Coloured communities of the two regions respectively, while Krugersdorp and Benoni catered for the Indian community. Only in the 1970s and 1980s was this policy reversed and the smaller centres and nearly all the Witwatersrand towns provided with separate Indian and Coloured group areas.

Changes in the application of the Group Areas Act took place in the 1980s as constitutional developments necessitated a more generous approach to the problems of the Indian and Coloured communities. Thus there was a decline in the extent of the White proclamations; indeed, in the course of the 1980s more White areas were being deproclaimed than new areas proclaimed. However, there were notable extensions to the Indian and Coloured

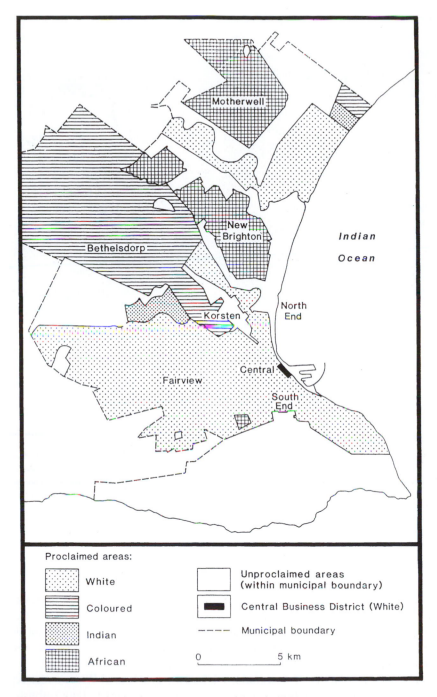

Motherwell

New
Brighton

Indian

Ocean

Bethelsdorp

North
End

Korsten

Central

Fairview

South
End

Proclaimed areas:

White

Coloured

Indian

African

Unproclaimed areas
(within municipal boundary)

Central Business District (White)

Municipal boundary

0   5 km

Figure 4.6 Port Elizabeth group areas proclaimed, 1991
Source: Based on information supplied by the Port Elizabeth Municipality

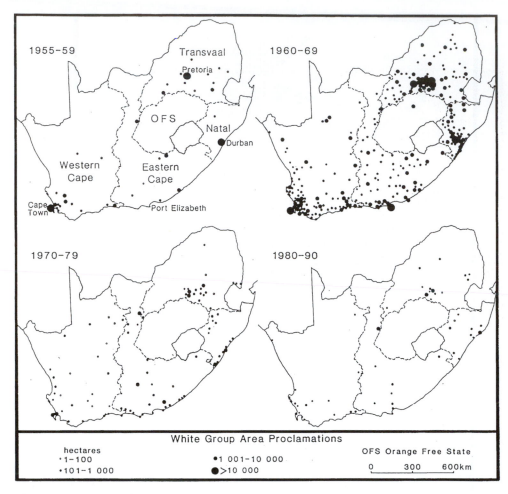

Figure 4.7 Sequence of White group area proclamation
Source: Based on information extracted from the files of the then Department of Planning, Provincial Affairs and National Housing, Pretoria

areas allowing for housing expansion. In some cases the extended areas came from depro-claimed White areas. In contrast, the Chinese were incorporated administratively into the White group in 1984, necessitating the inclusion of the Chinese group area into the White area. It should be noted that honorary White status did not include the franchise for the community.

Most significantly, a number of inner suburbs, initially proclaimed White, were repro-claimed for the communities originally resident in them. The most notable suburb was Woodstock in Cape Town, reproclaimed for the use of the Coloured population (Figure 4.10). Mayfair in Johannesburg was similarly treated for the Indian community. Thus the grand apartheid pattern of sectoral racial divisions of the city as originally conceived was interrupted by a number of non-conforming enclaves by the early 1980s.

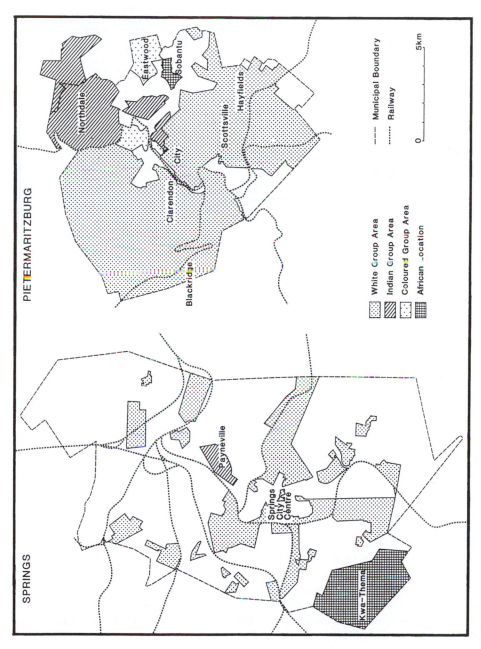

Figure 4.8 Contrasting patterns of proclamation, Springs and Pietermaritzburg
Source: Based on information extracted from the files of the then Department of Planning, Provincial Affairs and National Housing, Pretoria

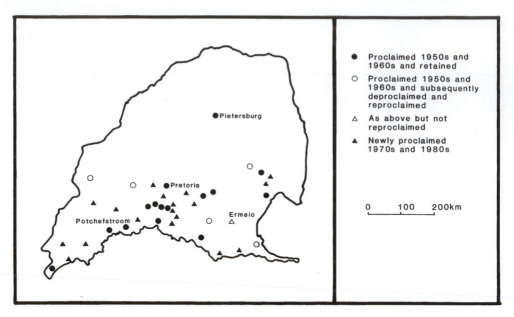

Figure 4.9 Centralization of Coloured areas in the Transvaal
Source: Based on information extracted from the files of the then Department of Planning, Provincial Affairs and National Housing, Pretoria

POPULATION RESETTLEMENT

The administration of the Group Areas Act was limited until the later 1950s by the lack of clear and practical procedures for converting the official intention into reality. However, in 1955 the Group Areas Development Board, later the Community Development Board, was given the power to purchase, sell and develop land, with extensive powers of expropriation. Accordingly, in the late 1950s and the 1960s surveys of disqualified people and affected properties were undertaken, to assess the extent of property expropriation and probable population movement involved in the execution of the schemes to fit the population to the newly drawn group areas boundaries (Figures 4.11 and 4.12). The detailed house-by-house survey of mixed areas provides one of the most comprehensive indicators of the extent of the racial integration which still survived in the late 1950s and early 1960s (Nel 1988).

It is not possible to establish accurate figures for the number of people displaced under the Group Areas Act. The use of the Slums Act and other measures to attain the ends of the Group Areas Act makes official statistics appear much smaller than they were in reality. However, the conservative figures produced by the authorities suggest that in the period up to 1984, when the administration of the Act underwent a major reorganization, a total of 126,000 families were displaced under the Group Areas Act. Of these, only 2 per cent were White, such was the skill of the Group Areas Board in drawing the boundaries to interfere with the White community as little as possible. By contrast, in many towns virtually the entire Indian and Coloured populations were displaced. This was particularly evident where scattered pre-1950 municipal housing estates were rezoned for White occupation.

The extent of the reorganization of the population distribution patterns was remarkable.

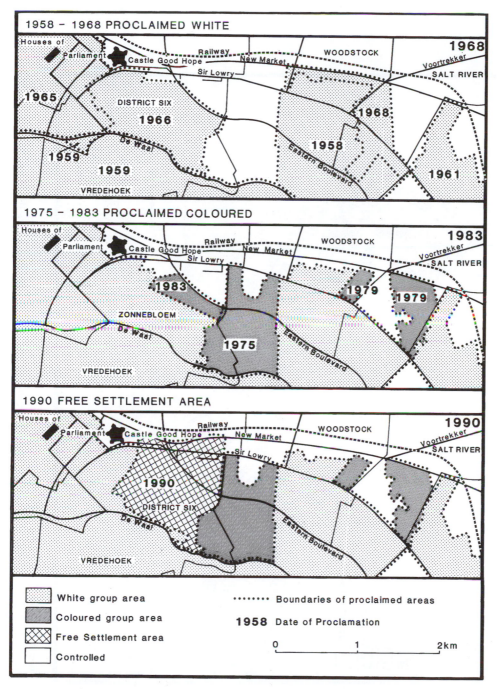

Figure 4.10 Changes to inner area zonings, Cape Town
Source: Based on maps supplied by the Cape Town Municipality

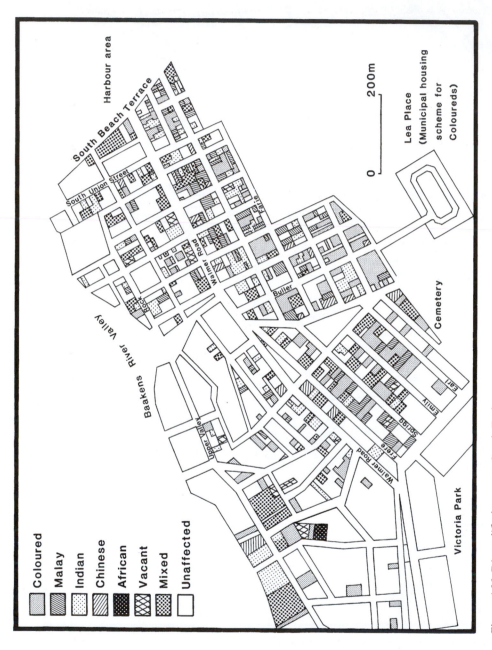

Figure 4.11 Disqualified persons, South End
Source: After J.G. Nel (1988) *Die Geografiese Impak van die Wet op Groepsgebiede en Verwante Wetgewing op Port Elizabeth,*
Port Elizabeth: University of Port Elizabeth

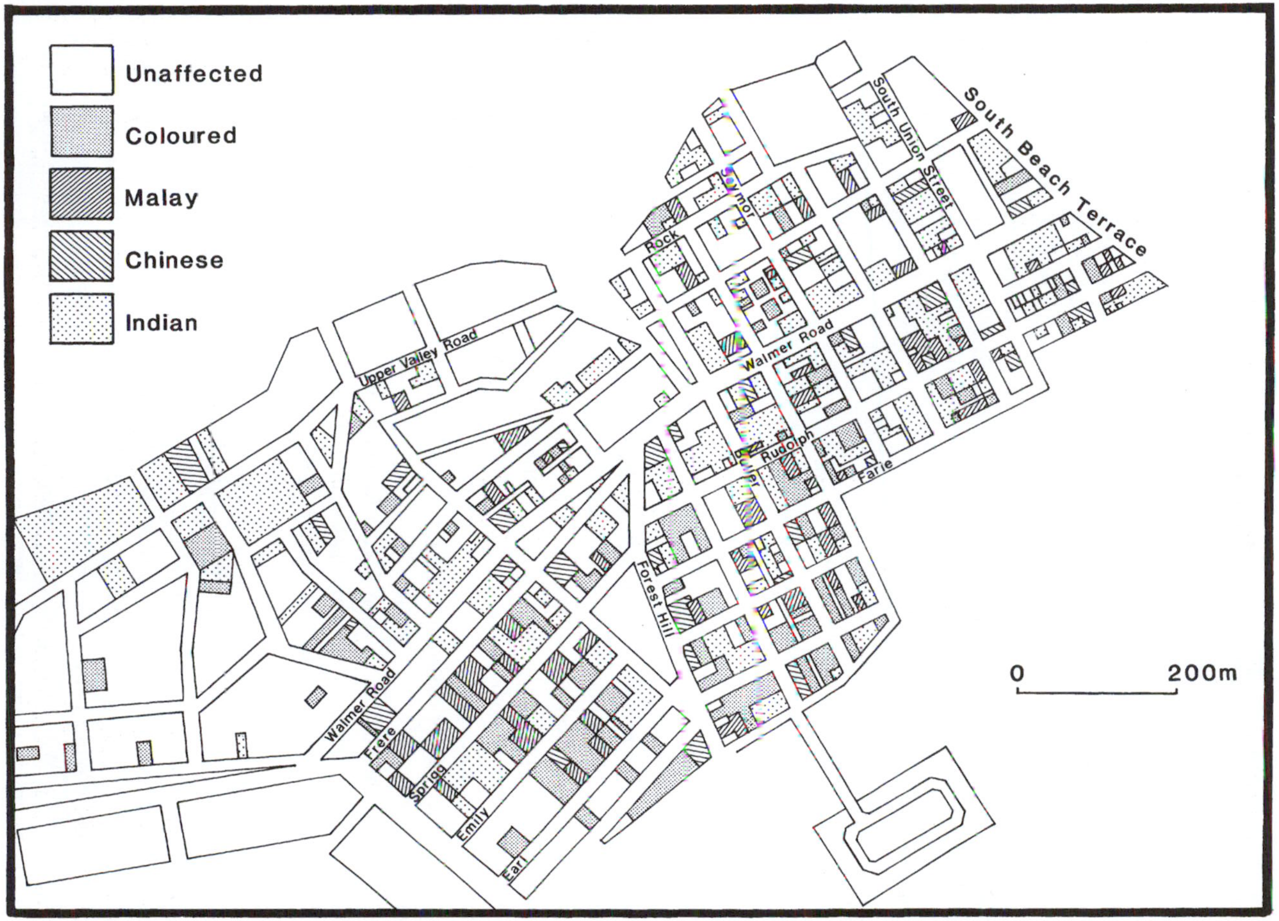
Unaffected
Coloured
Malay
Chinese
Indian
South Beach Terrace
South Union Street
Seymor
Rock
Walmer Road
Rudolph
Farie
Upper Valley Road
Forest Hill
Walmer Road
Frere
Sprigg
Emily
Earl
0
200m

Figure 4.12 Affected properties, South End
Source: After J.G. Nel (1988) *Die Geografiese Impak van die Wet op Groepsgebiede en Verwante Wetgewing op Port Elizabeth*, Port Elizabeth: University of Port Elizabeth

After 1950 most municipalities were forced, through central government sanctions, to operate as though effective group areas had been proclaimed, so far as the provision of new state housing was concerned. Newcomers migrating to the towns were directed to racially homogeneous suburbs, while families seeking their own homes for the first time were similarly directed. Forced migration from the central suburbs took place in the 1960s and early 1970s, parallel with expropriation and demolition.

The transformation of the patterns of population distribution is striking. In central Cape Town the contrast between the distribution recorded in the 1951 and 1985 censuses at census enumeration tract level, in consequence of the demolition of District Six and the expulsion of many other people as a result of the Group Areas Act, gives some indication of the detailed workings of the system (Figure 4.13). At a city-wide scale of suburbs, the transformation of the distribution of the population of Port Elizabeth between 1960 and 1991 indicates the extent of resettlement and the essentially peripheral nature of the new residential areas constructed for those people not classified as White (Figures 4.14 and 4.15).

AFRICAN RESETTLEMENT

The removal of the African population living in the central White designated urban areas was addressed in the 1950s under a battery of measures passed by previous administrations and through the tightening of residential control legislation effected by the euphemistically named Abolition of Passes Act (1951). The Natives Resettlement Act of 1954 provided for the removal of those owners and tenants with legal rights in urban freehold areas (Figure 4.16). Inner African-owned suburbs or individual plots were expropriated and the population resettled on the urban periphery or, where possible, in the homelands. The destruction of Sophiatown in Johannesburg is the most notorious of these removals. The inner African locations were also demolished or transferred to either Indian or Coloured occupation (Pirie and Hart 1985). A few, such as Wattville at Benoni, survived, although threatened with destruction for over forty years. The result of these policies was a substantial movement of population numbering approximately 750,000 people to new African townships on the periphery of the cities.

New legislation, the Natives (Urban Areas) Amendment Act of 1955, was introduced to remove such concentrations of Africans as the domestic servants in central city blocks of flats. Dr Verwoerd suggested that the arbitrary number of five African servants per block of flats was a 'wise' number to prevent overcrowding in the growing flatlands of Johannesburg and other cities (Mather 1987). The number involved in this particular movement in central Johannesburg was approximately 10,000, most of whom were rehoused in single-sex hostels in Soweto. The government then directed its attention towards other Africans living in the White suburbs. The Bantu Laws Amendment Act of 1963 limited the numbers of resident domestic servants to one per household, resulting in a further decline in the number of Africans living in the White suburbs (Hart 1976). However, proposals to remove all African servants, as part of a 'White by night' campaign, met with sustained resistance from White householders wishing to maintain their lifestyles.

The majority of the African townships were inherited from the previous era. The Natives (Urban Areas) Consolidation Act of 1945 with its various amendments provided the basic framework for African township control until the implementation of the Black Communities Development Act of 1984. It is worth noting that the area set aside for African townships

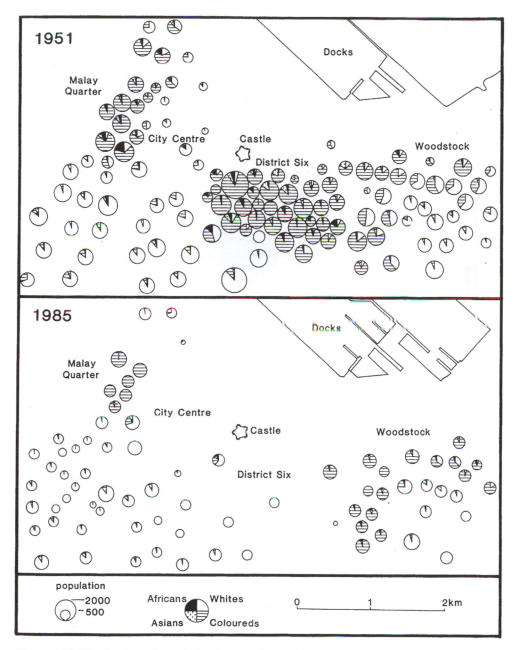

Figure 4.13 Distribution of population in central Cape Town, 1951 and 1985
Source: Based on information then held by the Central Statistical Services, Pretoria

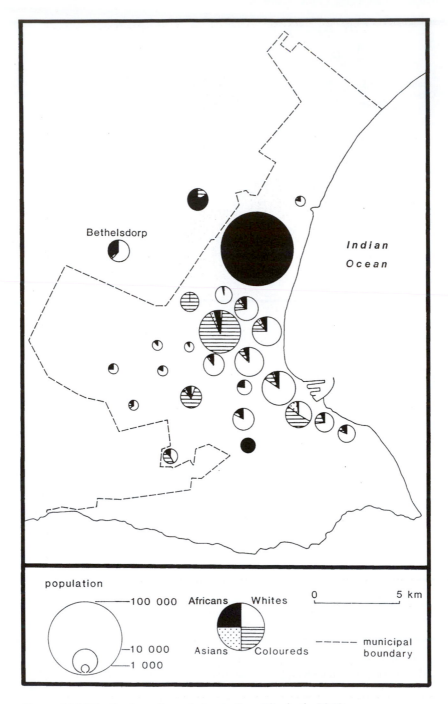

Figure 4.14 Distribution of population in Port Elizabeth, 1960
Source: Based on information then held by the Central Statistical Services, Pretoria

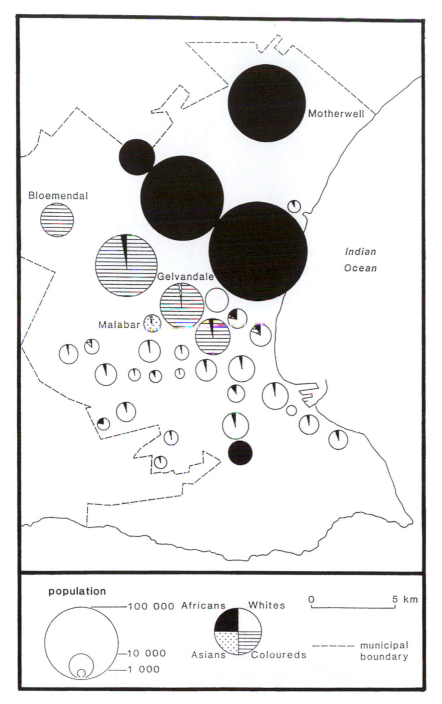

Motherwell

Bloemendal

Indian
Ocean

Gelvandale

Malabar

population

⌐100 000 Africans   Whites

0 ⌐⌐⌐⌐⌐ 5 km

⌐10 000   Asians   Coloureds

----- municipal
      boundary

⌐1 000

Figure 4.15 Distribution of population in Port Elizabeth, 1990
Source: Based on information then held by the Central Statistical Services, Pretoria

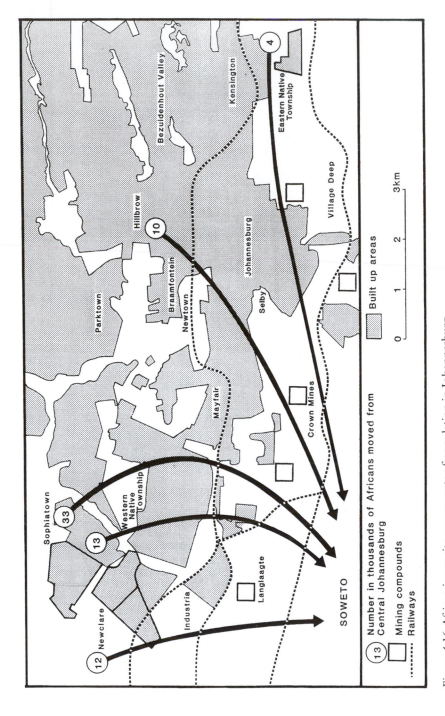

Figure 4.16 African inner city movements of population in Johannesburg
Source: Based on 1951 and 1960 census enumeration subdistrict information then held by the Central Statistical Services, Pretoria

until the late 1980s was small. This reflected the government's intention that the urban African population was, with few exceptions, temporary. The development of the homeland states, furthermore, was ostensibly to provide an incentive for Africans to live there and so depart from the White areas. This had the implications that only minimal housing was provided and private building was discouraged, in favour of homeland development. Indeed, Dr Verwoerd predicted that by the 1970s African migration to the cities would be reversed as the homelands would provide a greater attraction for work seekers (Davenport 1991: 270). Government policies were pursued with this unrealistic goal in mind.

However, the movement of people from the rural areas, and the quest for more urban living space, remained a constant theme of the period after 1948. In 1955 the Eiselen line was drawn around the western Cape, and subsequently extended, in order to declare the region a Coloured Labour Preference Area (Figure 4.17). Under this policy only Coloured labour was to be employed in the region, unless it could be shown that there were no suitable workers available. The Coloured rural reserves, notably in the arid district of Namaqualand, were occasionally referred to as the basis of a Coloured homeland, but no such entity ever entered official policy.

Accordingly, the official intention behind the Eiselen line was to remove virtually all Africans from the region. This policy was again a failure but was only recognized as such in 1983 after nearly thirty years of futile implementation. The policy resulted in the emergence of a serious African housing crisis in the Western Cape as the construction of state housing for the community ceased. Consequently African migrants were forced to squat illegally in settlements such as Crossroads on the periphery of Cape Town (Cook 1986). The official response was to demolish the shacks and send the people back to the homelands. In 1980 alone some 16,000 people were arrested for contravening the strict influx control regulations.

It was only with the lifting of influx control in 1986 that a massive reversal of this restrictive policy became evident. The provincial authorities then responsible for the designation of Black Development Areas sought out land on a substantial scale. In 1990 alone some 17,000 hectares were designated as Black Development Areas, compared with 124,000 hectares in the previous ninety years. During the final five months of zoning a further 6,000 hectares were designated, including six areas proclaimed on the last working day before the repeal of the racial zoning legislation (*Government Gazette*, 28 June 1991). No better illustration of the perseverance of the bureaucratic machinery could be presented. Black Development Areas remained remarkably crowded with densities on average twice that of the Coloured or Asian group areas, and eight times that of the White group areas.

AFRICAN ETHNIC SEGREGATION

The policy of state partition intruded into the planning of the apartheid city as the African population was not officially regarded as one entity. Ethno-linguistic differences noted for the purposes of establishing separate nation-states were emphasized in order to link the urban African population to their homelands, where they were to obtain political rights. In 1954 the government instituted the policy of ethno-linguistic segregation to 'assure the simplified and improved education of children in house [sic] languages; to maintain tribal discipline; to assist the efficient functioning of Bantu (African) authorities; to simplify municipal control; and to make for more harmonious living among the Bantu' (de Swardt 1970: 467).

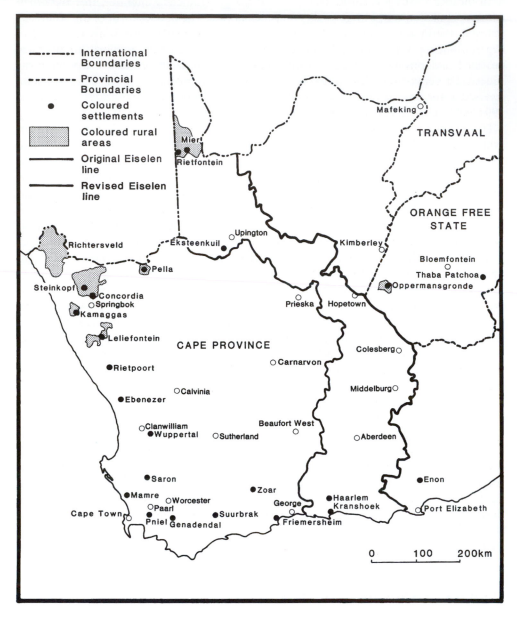

Figure 4.17 Eiselen line and Coloured settlements
Source: Modified after A. Lemon (1976) *Apartheid: A Geography of Separation*, Farnborough: Saxon House, G. Maasdorp (1980) 'Forms of partition', pp. 107–46 in R.I. Rotberg and J. Barratt (eds) *Conflict and Compromise in South Africa*, Cape Town: David Philip, and information supplied by the Provincial Administration of the Orange Free State

Accordingly, the African areas of towns were zoned for the different linguistic groups. In the majority of cities only one such group was present in any numbers. Thus the African townships of Durban were over 90 per cent Zulu-speaking, while the African population of Port Elizabeth was 97 per cent Xhosa-speaking. In such circumstances there was little to be achieved in segregation between different linguistic groups. However, on the Witwatersrand the African population had been drawn from all over South Africa, and no such dominance by one group was evident (Figure 4.18). In 1985 some 26 per cent of the population of the Witwatersrand African townships were Zulu-speaking and 15 per cent Tswana-speaking (Christopher 1989). The Xhosa, North Sotho and South Sotho also each exceeded 10 per cent of the total Witwatersrand townships population, although local concentrations varied substantially.

Segregation was achieved through the manipulation of housing allocation systems for new arrivals in the African townships. The policy was implemented in varying ways and without the determination evident in the enforcement of group area policies. In the older inner townships no zoning was possible as housing was allocated on an 'ethnic-blind' manner according to the availability of housing and progress in reducing the housing list. In expanding townships, the new extensions alone could be zoned, leaving some ethnically integrated and others segregated (Morris 1980). Only in completely new townships planned and settled after 1954 could the policy be enforced with some degree of completeness. Owing to the magnitude of the housing programme and the speed at which resettlement from the remainder of the cities took place, housing managers were more usually concerned with supplying houses to those in need than observing the ethno-linguistic segregation policies of the government. Thus the majority of townships were zoned on a simpler three-fold Nguni, Sotho and 'Other' basis. Only in Daveyton was an attempt made to segregate each group from virtually every other (Figure 4.19).

As a measure of the 'success' of the policy it should be noted that in the older townships, which experienced no expansion after 1954, ethnic segregation levels were remarkably low, usually signifying little more than random distributions. In most of the new and expanded townships segregation between the major groupings was substantially higher, suggesting a degree of success in implementing the policy. However, only in Daveyton did groups such as the Xhosa and Zulu experience segregation; elsewhere they were integrated with apparently random distributions relative to one another. Nowhere were intra-African indices of segregation as high as those between the White and African populations.

RESIDENTIAL SEGREGATION LEVELS

One of the basic questions confronting an assessment of urban apartheid is the degree to which the official plan was converted into reality. For an examination of the levels of residential segregation attained by apartheid social engineering, the populations enumerated at tract or subdistrict level at the time of the 1991 census have been analysed. The results of the 1991 census have serious problems attached to them due to survey sample methodology and the policy of non-cooperation adopted by the African National Congress during a period of political strife. Furthermore, by 1991 the results of the easing of several apartheid laws were beginning to be noticeable in White group areas. The census of 1991, despite its imperfections, may be regarded as a measure of the apartheid extreme.

The marked regional variations in segregation levels noted in the 1951 census remained in evidence, with a few exceptions, but all indices were raised to remarkably high levels. The full

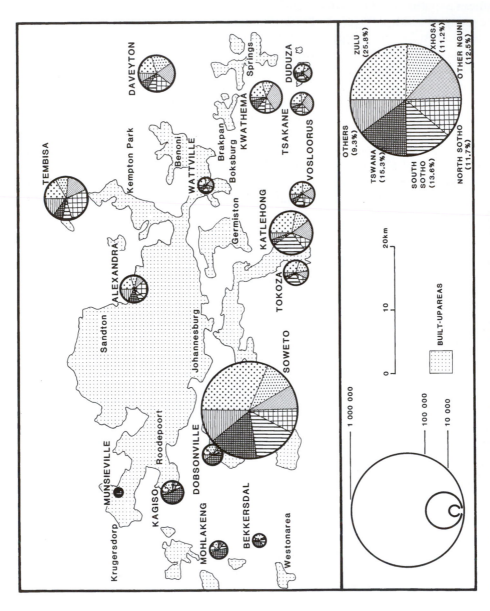

Figure 4.18 Ethnic composition of Witwatersrand African townships, 1985
Source: Based on information then held by the Central Statistical Services, Pretoria

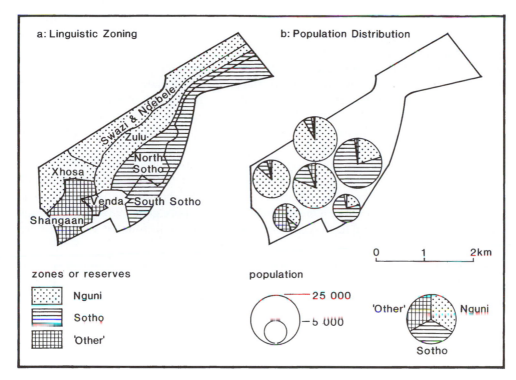

a: Linguistic Zoning

b: Population Distribution

Swazi & Ndebele

Zulu

North Sotho

Xhosa

Venda · South Sotho

Shangaan

zones or reserves

Nguni

Sotho

'Other'

population

25 000

5 000

'Other'

Nguni

Sotho

0   1   2km

Figure 4.19 Ethnic zoning in Daveyton
Source: Based on South Africa (1954) *Planning Residential Areas for Bantu in Urban Areas: Ethnic Grouping,* Pretoria: Department of Native Affairs, and 1985 census enumerators' returns then held by the Central Statistical Services, Pretoria

impact of apartheid planning was experienced in the Western Cape, which had exhibited the greatest degree of inter-racial residential integration before the implementation of the Group Areas Act. By 1991 it had become one of the most highly segregated regions in the country. The patterns which emerged were those of almost total segregation. Only 8.6 per cent of the urban population of South Africa (excluding the homelands) lived outside their designated areas by 1991. Significantly, only 0.2 per cent of the White population lived outside the White group areas, whereas 10.5 per cent of the Coloured population, 12.0 per cent of the Asian and 12.8 per cent of the African population did so (the latter figure is reduced to under 10 per cent if the urban population of the homelands is considered). Most of those persons lived in segregated areas zoned for another group, notably Africans in old central locations due for demolition or in barrack accommodation adjacent to industrial sites. Others not conforming to the pattern included the declining numbers of domestic servants living on the properties of their employers. However, despite forty-one years of official attempts to reduce the number of African domestic servants, few census tracts in White towns failed to record at least 5 per cent of the population as belonging to other groups.

Nevertheless, marked variations occurred within South Africa. The White populations of the towns of the Orange Free State and the Eastern Cape were significantly more segregated than those in Natal and the Transvaal. This state of affairs parallels the colonial and early

Union pattern closely, albeit at much higher levels of separation. Remarkably, there was no significant difference in segregation levels between large and small towns (Figure 4.20).

The Coloured population had undergone an exceptional transformation as a result of apartheid urban planning. Both the Western and Eastern Cape were significantly more segregated than either Natal or the Orange Free State by 1991 (Figure 4.21). Again the nineteenth-century heritage is evident in that the Orange Free State system of housing the Coloured population in the African locations was only undergoing slow change by 1991, while the Natal Coloured community had traditionally been more closely integrated with the White population. However, the dramatic transformation of Cape towns, notably those in the Western Cape, is most evident.

The Asian population had been subjected to close regulation since initial immigration in the nineteenth century. Thus segregation levels were high, but with significant regional variations, reflecting the colonial patterns (Figure 4.22). The absence of Indian communities in the Orange Free State is immediately apparent, although in 1984 the restrictions on residence in the province had been lifted. Elsewhere the Indians in the Transvaal were

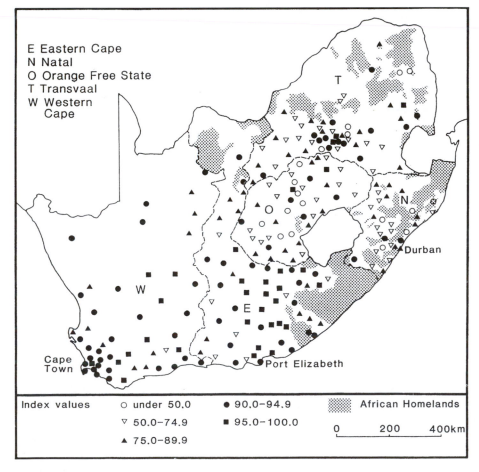

Figure 4.20 White index of segregation, 1991
Source: Indices calculated from the census enumerators' returns then held by the Central Statistical Services, Pretoria

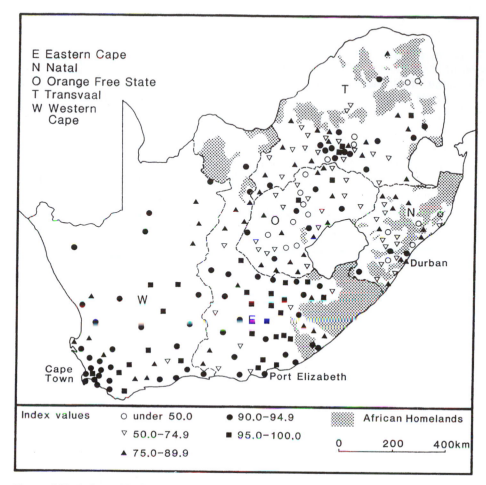

E Eastern Cape
N Natal
O Orange Free State
T Transvaal
W Western
  Cape

Index values		
○ under 50.0	● 90.0–94.9	▓ African Homelands
▽ 50.0–74.9	■ 95.0–100.0	
▲ 75.0–89.9		0 200 400km

Figure 4.21 Coloured index of segregation, 1991
Source: Indices calculated from the census enumerators' returns then held by the Central
Statistical Services, Pretoria

significantly more segregated while those in the Western Cape remained less segregated. In
the latter case the link through religion between the Cape Malays and Indian Moslems is
notable, often making clear-cut classifications impossible.

The pattern of African segregation is more complex as the effects of the incorporation of
African urban communities into the homelands need to be considered (Figure 4.23).
Accordingly, in Natal and parts of the Transvaal and eastern Cape Province the African
suburbs of a number of towns were included in the neighbouring homeland. The most
notable examples are Durban and East London where virtually the entire African suburban
areas were built in the homelands or were subsequently transferred to them. In other cases
the African population was rehoused at considerable distances from the original town and no
physical connection was evident. The construction of Botshabelo some 30 kilometres from
Bloemfontein is a particularly striking example, as some 500,000 people lived in the settlement by the late 1980s. Similar, if smaller, all-African towns were built elsewhere in the

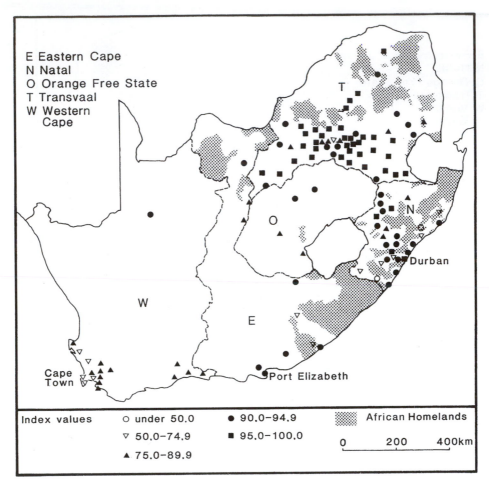

Figure 4.22 Asian index of segregation, 1991
Source: Indices calculated from the census enumerators' returns then held by the Central
Statistical Services, Pretoria

homelands to serve a distant White town. Thus the significantly lower segregation levels
particularly noted in Natal towns must be regarded with caution. However, the significantly
higher levels recorded in the Orange Free State again reflect the colonial pattern.

The administrators of apartheid planning consequently had achieved virtually total segregation in residential patterns in most South African cities by the 1980s. Indeed, segregation
levels by the time of the 1970 census indicate that in the majority of cities implementation of
segregation was nearly complete (Christopher 1990). This suggests that by the early 1990s
remarkably few urban dwellers had lived even part of their adult lives in racially and ethnically integrated conditions, and the process of dismantling apartheid offered unknown challenges to those involved. In this respect apartheid urban planners may be regarded as having
succeeded in their aims by the 1970s. However, pressures in the 1980s led to modifications
of this rigid model.

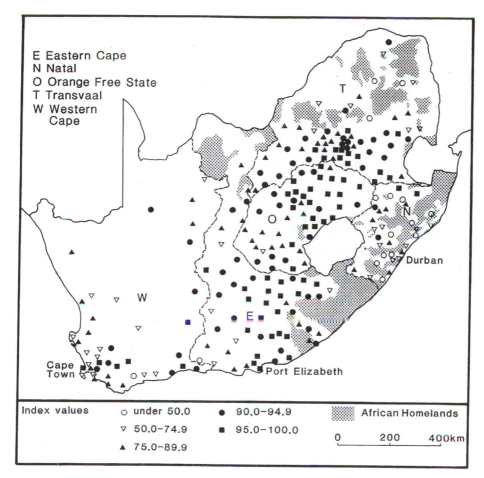

E Eastern Cape
N Natal
O Orange Free State
T Transvaal
W Western Cape

Index values		
o under 50.0	● 90.0–94.9	African Homelands
▽ 50.0–74.9	■ 95.0–100.0	
▲ 75.0–89.9		0 200 400km

Figure 4.23 African index of segregation, 1991
Source: Indices calculated from the census enumerators' returns then held by the Central Statistical Services, Pretoria

BUSINESS AREAS

The Group Areas Act was concerned not only with residence but also with business activities. Thus, in continuation with the flow of pre-1948 anti-Indian trading measures, one of the prime purposes of the Act was to remove Indian businessmen from White areas. The Group Areas Act specified disqualification from carrying on a business except in the group area of a member's racial classification. Accordingly, in the various central and other business districts trading was restricted to White people. In Natal the dual structure of central business districts was directly affected. In the majority of Transvaal towns a similar structure existed, although the Asiatic bazaars had been designated earlier. Elsewhere Indian traders occupied premises among other groups, although rarely in sections of the financial sector of the central business district.

The expropriation of Asian- and Coloured-owned businesses proceeded with the declaration of group areas. It was possible for shop owners to gain temporary exemption from the terms of the Act through the permit or appeal systems. However, the numbers surviving orders to cease business declined, and only in the late 1970s did Chinese traders begin to obtain permanent exemption from the Act in certain sections of the major cities. The principal area of survival was a section of the Indian central business district in Durban. The group areas proclamations disqualified the residents from living there but permitted trade to be continued (Figure 4.24). This was an unsatisfactory arrangement as the traders usually lived above the business premises. Furthermore, only a restricted part of the Indian trading area was allowed to survive, and not the fringe or adjacent residential areas. However, in 1984 the remaining residents were permitted to stay.

The apartheid ideal of traders being restricted to their respective group areas was clearly impossible for communities such as the Indians and Chinese who depended disproportionately upon trade for a livelihood. Suitable business locations were essential for traders who relied upon servicing the entire urban population. This was recognized in 1957 with the introduction of the concept of Free Trade Areas in which Whites, Coloureds and Asians could buy property and conduct business. Few such areas were ever developed. In Port Elizabeth the small Indian businesses displaced from other parts of the city were allowed to relocate on the edge of the industrial area. In Johannesburg the Indian businesses removed from the centrally situated Diagonal Street area were offered sites in a new Oriental Bazaar to the west of the central business district. Similar Oriental Bazaars were established in other cities.

In 1986, as part of the government's attempt to co-opt the Indian community into the new constitutional dispensation, restrictions on trading were relaxed. Free Trading Areas were proclaimed in the majority of central business districts (Bahr and Jurgens 1991). In the main, the areas involved were small, coinciding with the effective central business district, but excluding small suburban shopping areas which remained almost exclusively in White hands (Figure 4.25). However, the application of the regulations became progressively more flexible with the result that through the permit system other areas were opened to traders of other groups, prior to the repeal of all regulations in 1991.

It should be noted that although the authorities were anxious to remove Indian and other traders from White areas, there was little concern with other group areas, where Indian, Chinese and Coloured traders were allowed to continue with little interference. African areas were excluded as only African-owned businesses were permitted within the designated African areas. These were restricted as limits were placed upon the size of stores (initially 150 and then 350 square metres) and the range of goods (only daily convenience goods) that could be sold until the late 1970s. Only one licence could be issued to an individual and policy decisions decreed that successful businessmen be directed to the homelands. Thus few shopping areas in African suburbs developed to any size before the 1980s. Alcohol sales, for example, long remained a state monopoly in African areas, as a dubious source of revenue for the municipalities and subsequent Administration Boards. It had been a principle of the administration of the African townships since colonial times that no expense should fall upon the White ratepayers. The municipal and, later, state monopoly of alcohol sales provided the minimal revenues required to administer, but not improve, the townships.

FREE SETTLEMENT AREAS

A further relaxation of the original apartheid plan came into effect in 1990 as a result of the Free Settlement Areas Act passed two years earlier. The workings of the Group Areas

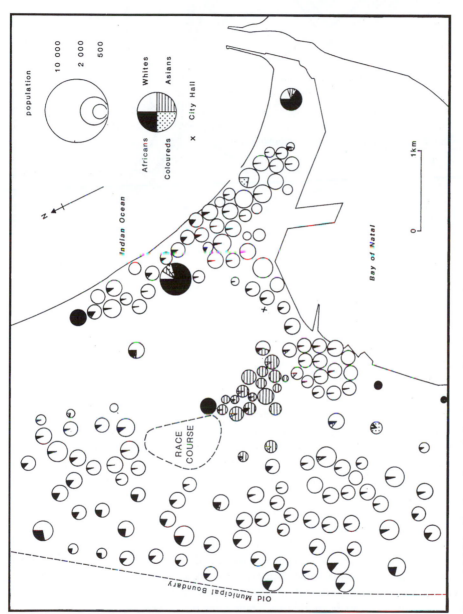

Figure 4.24 Distribution of population in central Durban, 1985
Source: Based on information then held by the Central Statistical Services, Pretoria

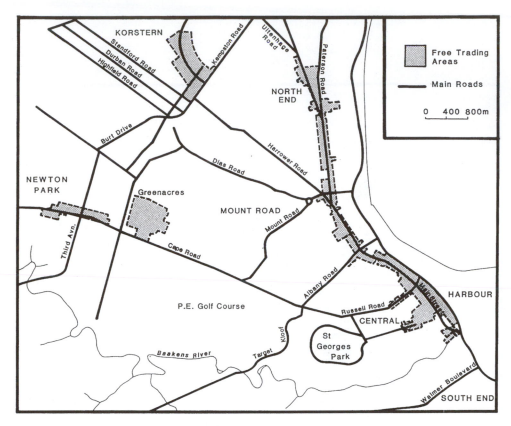

Figure 4.25 Free Trading Areas, Port Elizabeth, 1986
Source: Based on maps supplied by the Port Elizabeth Municipality

machinery were adversely affected by the decision of the Transvaal Supreme Court in the Govender case in 1982, when it was ruled that evictions under the Group Areas Act could only be effected if alternative accommodation could be offered (Morris 1994). In view of the chronic housing shortage in the Coloured, Indian and African residential areas, alternative accommodation was rarely available. Thus evictions under the Act virtually ceased. No amending legislation was introduced, as the government at the time wished to attract the Coloured and Indian communities into the new constitutional structures.

The result was an influx of Coloured, Indian and African people into the inner suburbs and a few other areas of the major cities (Hart 1989; Jurgens 1993; Rule 1989). The high-density flatlands attracted people seeking accessible and often single accommodation near the city centre and also wishing to flee from overcrowded housing in the peripheral townships (Figure 4.26). Other inner suburbs where some members of either the Indian or Coloured communities had been able to retain their properties also attracted returnees. Hence Mayfair in Johannesburg attracted Indian families, while elsewhere inner suburbs attracted a wide range of persons, including European immigrants. In this manner 'grey areas' were brought into being throughout the major centres of the country, reproducing, on a highly limited scale, the inner mixed suburbs of colonial times.

The government sought to control and limit the extent of this development and so passed

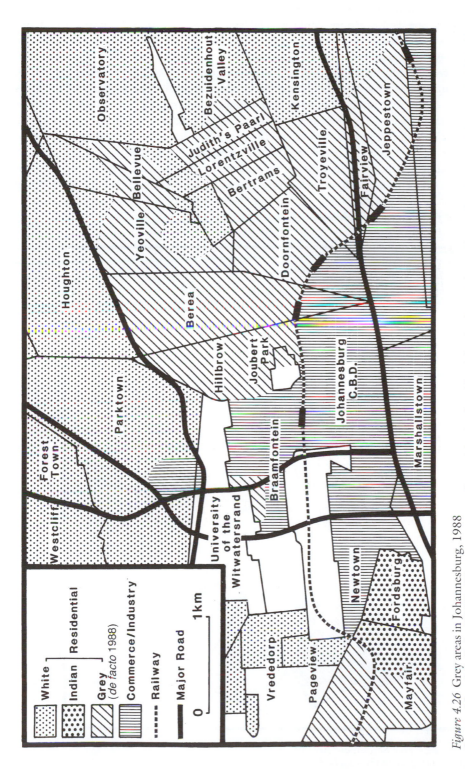

Figure 4.26 Grey areas in Johannesburg, 1988
Source: After S.P. Rule (1989) 'The emergence of a racial mixed residential suburb in Johannesburg: the demise of the apartheid city?',
Geographical Journal 155: 196–203

the Free Settlement Areas Act of 1988. The Act introduced the concept of areas where people of all population groups could live. A wide range of proposals, both official and unofficial, were put forward to implement the purposes of the Act. In a number of cases City Councils proposed that entire cities be opened to settlement by all, in others small new suburbs were offered, as though 'grey' represented another 'colour' on the group areas map, necessitating the establishment of separate local administration. The majority of the areas approved were for such undeveloped peripheral zones, where no one would be forced to live in an integrated society. The final area proclaimed at Uitenhage in 1991 was of this nature (Figure 4.27) Included in this section was the proclamation of the demolished section of District Six in Cape Town as a Free Settlement Area (see Figure 4.10). Few settled areas were proclaimed as the majority of inquiries into the objections were not completed when the Free Settlement Areas Board effectively ceased to function in February 1991. Only fourteen of the fifty-six formal applications for the establishment of Free Settlement Areas were taken through to proclamation.

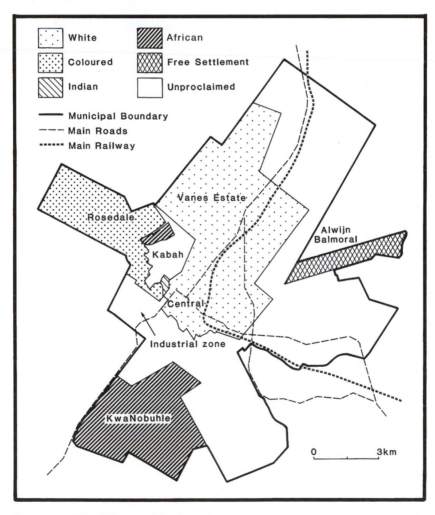

Figure 4.27 Free Settlement Area in Uitenhage
Source: Based on maps supplied by the Uitenhage Municipality

The gradual relaxation of the rules and regulations governing the apartheid city resulted in the emergence of the 'modernized' or late apartheid city (Lemon 1991a; Simon 1989; Smith 1992). The government of President P.W. Botha after his assumption of power in 1978 had sought to reform apartheid in order to make it more acceptable both internally and internationally. In the process, the details of group areas, separate amenities and urban administration had been modified. In many respects this was political accommodation to ensure the survival of ultimate White control. The broad plan of the apartheid city was firmly in place, with high degrees of compliance with its strictures (Figure 4.28). The creation of open residential areas was one of the most notable features. However, the recognition of African permanence in the cities through, first (1982), the ability to own houses but not land, and then (1986), recognition of freehold rights was a major philosophical departure from the initial apartheid model. Thus the construction of permanent high- and middle-class freehold housing marked a major change in the appearance of the African areas in the 1980s. However, taken together, these were relatively minor modifications to an urban entity firmly held in place by its physical construction. The emergence of persistent and growing areas of African squatting both within the African areas and on the margins of the cities was less acceptable, but beyond the capabilities of the government to prevent (Crankshaw and Hart 1990).

The incremental nature of the reform was characterized by the repeal of the Prohibition of Mixed Marriages Act and Section 16 of the Immorality Act in mid-1985, which permitted people of different races to marry or live together. However, the Group Areas Act remained in force, nominally preventing the couples from living in the same place. The legal problem was solved by the racial reclassification of one of the partners and movement to the appropriate group area. It might be noted that the option of reclassification to White was rarely permitted.

NEW TOWNSCAPES

The apartheid city was deliberately created by the South African government over a period of forty years. The broad zoning and governmental control over the detailed planning of the city is evident in a variety of aspects. Extensive new residential areas established by the government for the Coloured, Indian and African populations were built with houses of remarkable uniformity, only softened as tenants and later owners began the process of house improvement, through building new extensions, repainting and gardening. Nevertheless, many housing estates remained in appearance much as they had been initially, as the inhabitants were too poor to afford to improve or even maintain their properties. The appearance of uniformity, particularly in the African areas, was intensified by the speed with which they were built. Half of the formal housing of Soweto was constructed in a period of under twenty years from the early 1950s. Only in the 1980s were private developers permitted in the African areas, providing a range of housing types and environments. Even so, the street plans and plot sizes were still controlled by the authorities, intent on restricting the extent of the African areas. In contrast, the White extensions to the cities have followed the free market principles of American cities, with the extensive sprawl of smallholdings, unused farmland and patchy suburban development.

Within the inner core the apartheid city again followed the twin aspects of First World, capitalist, modern high-rise business districts, with surrounding zones of transition. However, strong evidence of Second World, socialist planning was present in most cities where

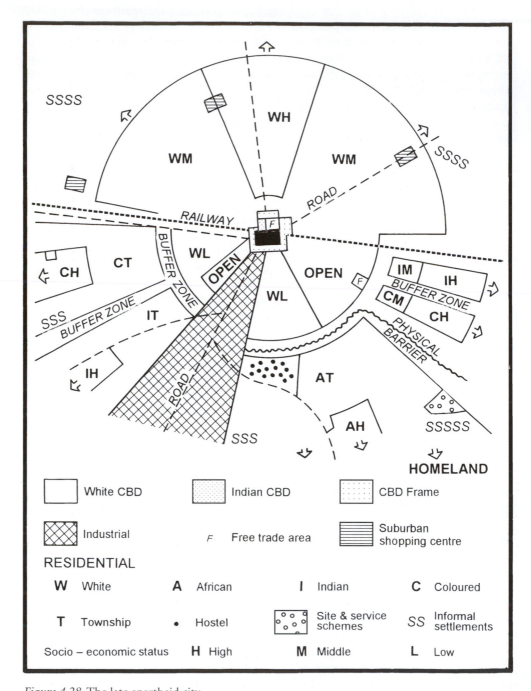

Figure 4.28 The late apartheid city
Source: After D. Simon (1989) 'Crisis and change in South Africa: implications for the apartheid city', *Transactions of the Institute of British Geographers* 14: 189–206

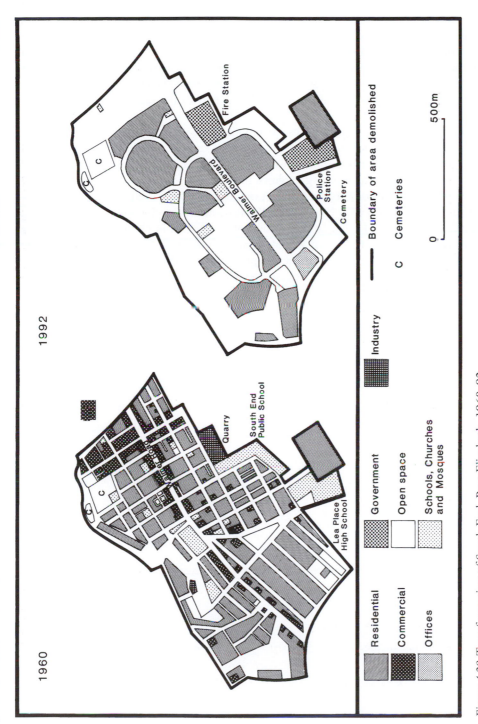

1960

1992

Residential | Government | Industry | — Boundary of area demolished

Commercial | Open space | C Cemeteries

Offices | Schools, Churches and Mosques

0 500m

Lea Place High School

Quarry

South End Public School

Warmer Boulevard

Fire Station

Police Station

Cemetery

Figure 4.29 Transformation of South End, Port Elizabeth, 1960–92

Source: Based on information extracted from *Donaldson's Port Elizabeth Directory 1959/60, 1959*, Port Elizabeth, and air photographs for 1960, and fieldwork, 1992

inner, integrated areas were demolished, providing the authorities with a reserve of land to be used for governmental purposes. The process predates the official adoption of apartheid in 1948. The civic centre at Kimberley was built on the site of the Malay Camp, which was demolished as a clearance scheme under the Slums Act. Expropriated land thus provided space for the construction of new government buildings close to city centres. In Cape Town and Durban new technikons were built on the extensive areas of demolished houses in District Six and Warwick Avenue respectively, while in Port Elizabeth a new police station and the main fire station were built in South End as part of the radical replanning of the suburb (Figure 4.29).

In a number of cases the land was subsequently used for White housing, resulting in the appearance of modern inner suburbs close to the city centre. The transformation of the predominantly African suburb of Sophiatown into the White suburb of Triomf was a particu-larly emotive example, as the renaming suggests. In other cities the cleared expropriated land was left derelict as no decision could be taken on its future which did not involve contro-versy. Thus Cato Manor in Durban remained as a wasteland in the middle of the metro-politan area (Maharaj 1994). The largest portion of District Six in Cape Town fell into this category as subsequent residential redevelopment schemes were considered to be inappropriate and the derelict area served as a political reminder of opposition to govern-ment policy. Elsewhere houses which were not demolished were resold to Whites as part of a 'gentrification' programme.

The enforced expropriation of property by the government for political purposes poses a substantial number of unanswered questions for any investigation of land rights in the post-apartheid era, as well as raising the whole question of the meaning of security of title.

5 Personal apartheid

Apartheid also operated at the micro level where separation affected the details of the daily life of the population. The policy was intended to eliminate virtually all personal contact between members of different population groups, except within the master–servant or employer–employee relationship, both of which were essentially viewed in White–Black terms. Attention was focused upon the physical isolation of the White group from the others. For most personal apartheid matters there were only two salient groups: Whites and 'Non-Whites'. Because of the racial dualism which was evident in personal apartheid, it is proposed periodically to refer to all who were not regarded as White as 'Non-Whites', however demeaning the term was considered to be. It illustrates the feeling of the intent of the multitude of laws and regulations issued to prevent those who were not classified as White from occupying and using declared White space. The term was widely used in everything from government notices to park benches.

Personal apartheid operated at various levels including the prevention of marriage or cohabitation between a member of the White group and a member of another group. It needs to be emphasized that the prohibition did not apply to intermarriage between members of the African, Coloured and Asian groups. In the occupation of space the laws operated from the level of separate park benches and entrances to buildings, to separate transport systems, schools, hospitals, and, ultimately, cemeteries.

A host of laws was introduced to enforce what was described as 'petty apartheid', as opposed to 'grand apartheid', involving political separation and urban residential segregation. Many of the laws had their predecessors in the pre-1948 era, but were made far more effective and comprehensive by the National Party government. Thus the Prohibition of Mixed Marriages Act (1949) and the Immorality Act Amendment Act (1950) were two of the earliest apartheid measures designed to preserve the imagined racial purity of the White group. Segregation was also enforced within the private domain. As a further measure, African and Coloured domestic servants living on White-owned properties were required to be provided with physically separate premises and entrances. Residential segregation was thus enforced within the confines of a single residential plot (Figure 5.1).

Separation at the scale of individual buildings was tightened. This ranged from the provision of separate counters in post offices to separate entrances to public buildings. Even separate stairways and lifts in multi-storey buildings were instituted where possible. Separate restrooms were also deemed essential. Such requirements had to be taken into account in the design of public and increasingly also of private corporate buildings. Thus the White and Non-White experiences of the same building were very different, as often separate offices dealt with the needs of the two groups. The plan of the Steytlerville post office illustrates the resultant duplication of entrances and public counters (Figure 5.2) (Herholdt 1986).

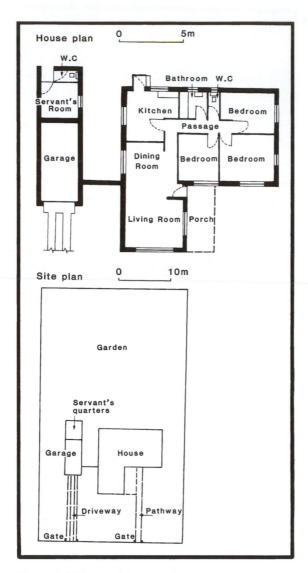

Figure 5.1 House with servant's quarters
Source: Based on building plans supplied by the Port Elizabeth Municipality

The excesses of personal apartheid may be illustrated by the experiences of Indian, Coloured and African students and doctors at the University of the Witwatersrand Medical School and the Johannesburg General Hospital after 1948 (Browde *et al.* 1998). Non-White students were not allowed to come into contact with White patients. Only in 1987 were they allowed into the White sections of the hospital, but not in maternity and gynaecology wards (finally opened in 1990). Most ludicrous, they were forbidden to wear white coats or carry stethoscopes while in the hospital, thereby symbolically to deny their status in the eyes of White patients. Furthermore, they had to enter and leave the hospital by back entrances. The use of separate dining facilities was enforced, even to the extent that the crockery and utensils were marked NEDR (Non-European Dining Room) to distinguish

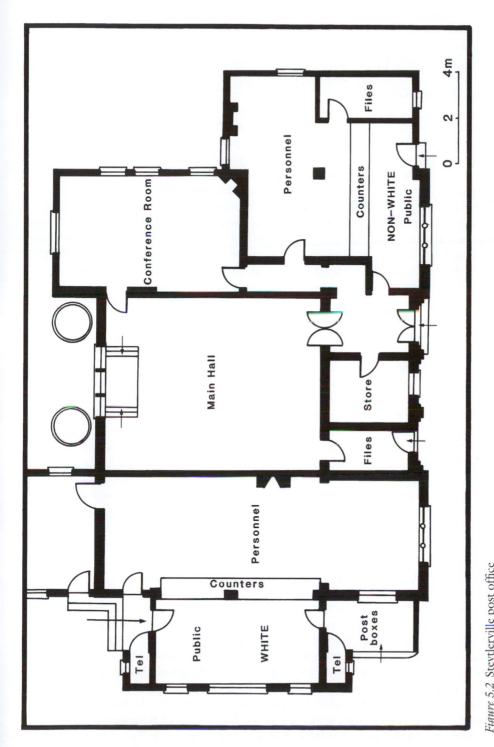

Figure 5.2 Steytlerville post office
Source: Modified after A.D. Herholdt (1986) *Die Argitektuur van Steytlerville met Riglyne vir Bewaring en Ontwikkeling,* Port Elizabeth: Institute of South African Architects

them from those in use in the EDR (European Dining Room). Their experience of training was thus very different from that of their White counterparts, although the final-year graduation photograph was traditionally taken outside the main entrance of the hospital, which they could not use.

SEPARATE AMENITIES

The Reservation of Separate Amenities Act in 1953 together with its numerous amendments sought to create separate social environments for the White and other population groups. The Act was closely linked with the racial land zoning of the Group Areas Act, in order to determine who might use particular facilities. Accordingly, parks, places of entertainment, and dining and hotel accommodation were all regulated and reserved for a single group. Hence it became impossible for a White person to drink a cup of tea with a Non-White person in any café unless they obtained a permit to do so, which until the 1980s was unlikely to be issued. The grandiose term, 'International Hotel', was introduced in 1975 to allow Non-Whites with sufficient financial means to stay in selected, generally expensive, hotels. The cheaper establishments remained segregated. Regulations were such that within the hotels granted 'international' status, swimming pools and dance floors were initially denied to patrons who were not White, unless the swimmer or dancer could produce a foreign passport; and even then a Transkeian or Bophuthatswanan passport was not acceptable. In Durban it might be noted that the majority of the International Hotels were situated opposite a 'Whites-only' beach, thereby leading to obvious disadvantages for African, Indian or Coloured patrons (Figure 5.3).

'Separate' under the Reservation of Separate Amenities Act did not mean equal. Where, for example, only one waiting room existed on a railway platform, the stationmaster was empowered to set it aside for the use of White people only. Only later might a second, Non-White, waiting room be built, at the other end of the platform.

Possibly one of the most controversial issues was the question of beach apartheid, as most of the coast line fell outside proclaimed group areas (Gibbon 1977). Owing to the problems of defining the space involved, the matter was only brought within the ambit of the Reservation of Separate Amenities Act by an amendment in 1960, which permitted local authorities to reserve both beaches and the adjacent sea for the exclusive use of a single group. The resultant zoning was remarkably complex and inequitable. Although in the majority of resorts the main beaches were reserved for Whites, peripheral beaches were reserved for other groups. Hence in Port Elizabeth the key sandy beaches south of the harbour were reserved for Whites and only those deemed less desirable, as adjacent to the industrial areas, too rocky, or too remote, were zoned for the other groups (Figure 5.4). For this purpose the full classification of the Group Areas Act was adopted with the absurd anomaly of separate beaches for Whites, Coloureds, Chinese, Indians, Cape Malays and Africans. Local authorities which made such zonings were then under the obligation to construct changing and ablution facilities for each of the groups. However, it was assumed by local town planners that only White people wished to enjoy, rather than could afford, beach recreation and hence few amenities were provided for the other groups until the 1980s.

Segregation was applied to sports clubs, sports grounds, swimming pools and other public open spaces. Sports facilities were governed by the Reservation of Separate Amenities Act, even if privately owned. These included such major users of land as golf courses which, even if not within a proclaimed group area, were governed by the race of the owner. In this

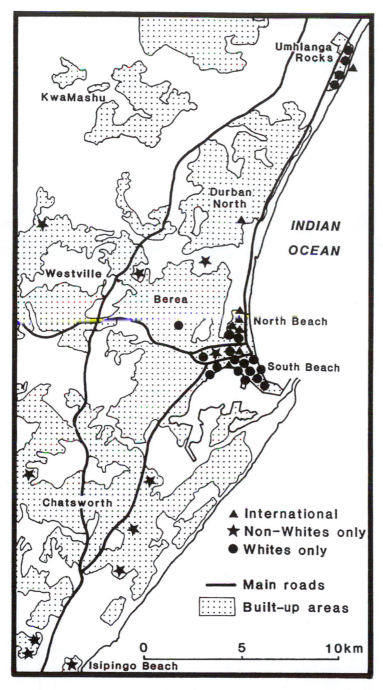

Within the map:

KwaMashu

Umhlanga Rocks

Durban North

INDIAN OCEAN

Westville

Berea

North Beach

South Beach

Chatsworth

Isipingo Beach

▲ International
★ Non-Whites only
● Whites only

——— Main roads
:::::: Built-up areas

0   5   10km

Figure 5.3 Hotel classification in Durban, 1985
Source: Based on information supplied by the Automobile Association of South Africa

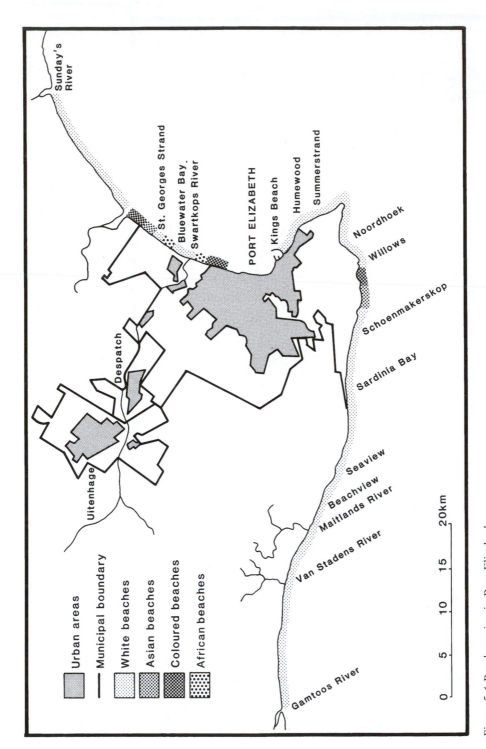

Legend:

- Urban areas
- Municipal boundary
- White beaches
- Asian beaches
- Coloured beaches
- African beaches

Labels on map:

Sunday's River, St. Georges Strand, Bluewater Bay, Swartkops River, PORT ELIZABETH, Kings Beach, Humewood, Summerstrand, Noordhoek, Willows, Schoenmakerskop, Sardinia Bay, Despatch, Uitenhage, Seaview, Beachview, Maitlands River, Van Stadens River, Gamtoos River

Scale: 0 5 10 15 20km

Figure 5.4 Beach zoning in Port Elizabeth
Source: After A.D. Gibbon (1977) 'Outdoor recreation survey of the Port Elizabeth area', unpublished MA thesis, University of Port Elizabeth

manner Whites could not compete against members of another population group in any sporting activity. Although participation remained segregated, some relaxation of the law was possible at certain sports stadiums and at race courses, where segregated enclosures for White and Non-White spectators were provided.

RELIGION

The government sought to introduce separation into church congregations, recognizing the importance of the religious community in social integration. In many respects the Group Areas Act effectively broke up integrated church congregations as disqualified people were removed from living in the vicinity of the church. Some continued to attend their original place of worship, even if this involved travelling considerable distances, as a means of registering their protest. However, children upon growing up felt less attachment to the place from which their parents had been forcibly removed, and attended more convenient places of worship. Furthermore, often the churches were sold and demolished when the suburbs as a whole were destroyed. Others were left without congregations and eventually converted to other uses, as notably the Central Methodist Church which became the District Six Museum in Cape Town. Significantly, the mosques were not demolished, but remained as a constant visible reminder of the impact of government actions against the religious community. Indeed, the visible reminders remained a significant element in sustaining opposition to apartheid (Mattera 1987).

The politically dominant church was fragmented into four separate entities, and acted as a model for government plans. The original Dutch Reformed Church (Nederduits Gereformeerde Kerk) remained for Whites, while the Dutch Reformed Mission Church (Nederduits Gereformeerde Sending Kerk) catered for Coloureds, the Reformed Church in Africa for Indians, and the Dutch Reformed Church in Africa (Nederduits Gereformeerde Kerk in Afrika) for Africans. The distribution of the denomination's churches followed the group areas pattern with few exceptions (Figure 5.5).

Other denominations did not follow the same approach and the church–state confrontation over the issue resulted in the status quo being maintained. Thus the Anglican Church of the Province of Southern Africa remained nominally integrated, although congregations tended to reflect the segregation of the residential areas. It does, however, also include the Order of Ethiopia, which is exclusively African and operates outside the general parochial and diocesan system. The host of African separatist churches, which came into being in the nineteenth and twentieth centuries, represented a blending of African traditional and European missionary traditions, which could not be accommodated within the Eurocentric churches. Such separation is therefore rooted in colonial attitudes rather than apartheid.

Spatial segregation continued even in death. The ultimate aspect of separation in social contact was the maintenance of segregated cemeteries. The majority of cemeteries in the nineteenth and early twentieth centuries were segregated by religious denomination, often with sectors designated for individual churches. This allowed for some measure of legal population group integration where churches were open to everyone, but racial segregation where congregations were segregated. Areas within cemeteries not designated for the religious denominations were usually segregated with descriptions such as 'European Free Ground'. However, after 1950 new municipal cemeteries were zoned exclusively for the different population groups, where this was not already done (Christopher 1995a). In the main it was achieved in conformity with group areas zonings, although the older cemeteries

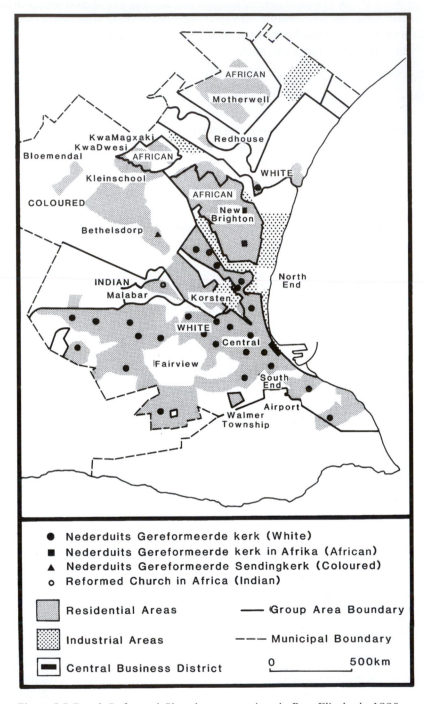

Figure 5.5 Dutch Reformed Church congregations in Port Elizabeth, 1990
Source: Based on information supplied by the Dutch Reformed Church

continued to be used where families had already purchased burial plots (Figure 5.6). The pattern of separate burials is likely to persist, with planning for separate burial grounds continuing after the repeal of the Reservation of Separate Amenities Act.

SEPARATE TRANSPORTATION

The enforcement of the Group Areas Act and the Natives (Urban Areas) Act ensured that movement between place of work and place of residence was usually separate, as a result of the different destinations to which the various groups of workers travelled when going home. However, detailed configurations of bus routes and bus stops were carried over to the provision of separate sections of trains and buses for Whites and Non-Whites. In the railway system, trains were divided into White and Non-White sections, each separated by class,

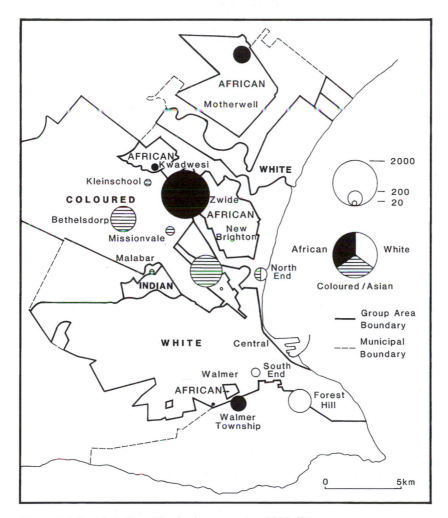

Figure 5.6 Burials in Port Elizabeth cemeteries, 1990/91
Source: Based on information supplied by the Port Elizabeth, Ibhayi and Motherwell municipalities and the Cape Provincial Administration

although third class was only provided for Non-Whites. This arrangement necessitated the provision of separate sections of railway platforms, each with an assemblage of ticket offices, waiting rooms and bridges over or tunnels under the tracks and an effective no man's land between the two.

Buses and trams presented a greater difficulty for the enforcement of segregation than trains, as everyone had to use a common entrance. Segregation had been imposed before 1948 in many services outside the Cape Province through the allocation of certain sections of a bus, usually the upper deck or the rear, to Non-White passengers and the remainder to Whites. In other centres separate buses and trams were run for Whites and Non-Whites, operating on different routes with separate bus stops and termini. In Durban even the buses were painted different colours for different groups to distinguish vehicles more readily. Some relaxation was locally permitted for Coloureds and Indians to use White facilities. In 1955 the government began to enforce regulations in those cities with integrated services, notably Cape Town and Port Elizabeth. At first (late 1950s) segregation within buses was imposed, with the reservation of separate sections for the two groups; later (late 1960s) separate buses were enforced (Figure 5.7) (Pirie 1990a). Bus segregation was shortlived as in the late 1970s it was relaxed and finally abolished even in those cities where it had been firmly entrenched before 1948.

SCHOOL EDUCATION

The provision of education had been remarkably segregated before 1948. However, the new government was committed to the institution of Christian national education, one of the

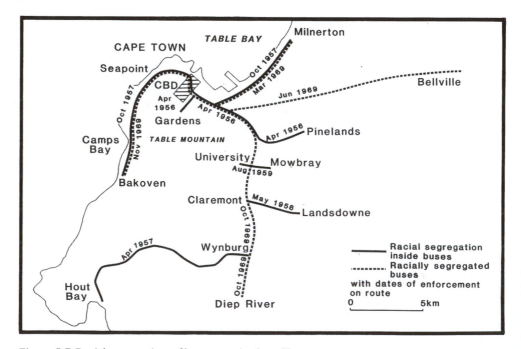

Figure 5.7 Racial segregation of bus routes in Cape Town
Source: After G.H. Pirie (1990) 'Racial segregation on public transport in South Africa 1877–1988', unpublished PhD thesis, University of the Witwatersrand

tenets of which was the promotion of ethnic identity and solidarity. In 1949 the government appointed a commission to inquire into the future of African education. The result was the formulation of a new separate state education system for the African population. It was also to be specific to African needs and not, in Dr Verwoerd's terms, for the creation of 'expectations in life which circumstances in South Africa do not allow to be fulfilled' (*Hansard* 1953: col. 3576). Under the new government Africans were not to aspire to certain positions in society and so education for such positions was not deemed necessary. Control over African education was taken by the Department of Native Affairs in 1954 under the Native Education Act, as the previous mission schools were regarded as fostering ideas, such as equality, which could not be encouraged.

One of the vital foundations of the new African education system was the provision of mother tongue instruction, at least at primary level beyond which few pupils progressed. The object was to fragment the African population into separate linguistic nations, each with an awareness of its own history and culture acquired through the education system. The requirement was the basis whereby different linguistic groups in the African suburbs were to be housed separately, and provided a means whereby control over urban education could be linked to the African homeland governments (Figure 5.8).

White schools were already separate, under the control of the four provincial authorities. Separate English- and Afrikaans-medium schools were provided throughout the country, together with such German schools as local communities could support. Dual- or parallel-medium schools were not encouraged, in the interests of maintaining Afrikaner cultural identity. Provision was made for separate Indian and Coloured schools within the provincial education systems. Between 1984 and 1994 these were administered by the relevant

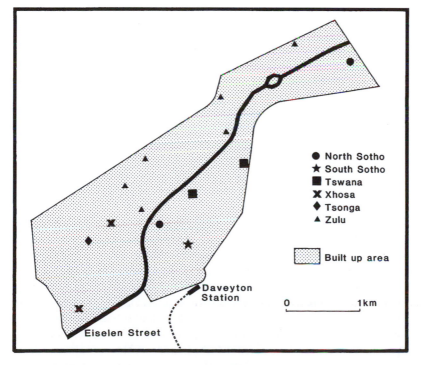

Figure 5.8 Primary school mother tongue, Daveyton
Source: Based on information supplied by the then Department of Education and Training

houses of parliament. Even exclusive Chinese schools were established in Port Elizabeth, in line with the Chinese group area. Only the private schools, notably those controlled by the churches, remained outside the direct control of the government. However, these schools were effectively run until the 1980s as segregated entities in order to obtain state recognition and accreditation. The number of pupils attending such schools was small in comparison with the total. Thus in the school sphere segregation was almost complete, exceeding that in residence (Figure 5.9).

In view of the significance of education in fostering the overall apartheid system, the control of African education in the homelands was passed to the homeland governments. Accordingly by the late 1980s there were some seventeen departments of education in the country, in addition to a central national ministry.

UNIVERSITY EDUCATION

University education was similarly segregated. The Extension of University Education Act (1959) provided for the establishment of a series of new ethnically based institutions for Africans, together with separate universities for Coloured and Indian students, despite strong protests from the established universities. In seeking to create the new institutions, the government indicated that 'they must be the bearers of their own national culture to stimulate that culture amongst their own national group' and that 'the future leaders should be educated and trained there, not to break down the colour bar but to retain it in the best interests of both Whites and non-Whites' (*Hansard* 1959: col. 6221).

In 1958 only 1,402 Asian, African and Coloured students (17 per cent of the total) had been registered at university institutions classified as White (Figure 5.10). Thereafter only Whites were to be admitted to the existing universities, with the exception of Fort Hare, which was taken over by the government as an institution for the Xhosa population. The only other exceptions were the medical school of the University of Natal, which was open to Asian, African and Coloured students only, and the correspondence institution, the University of South Africa. A few students who wished to pursue a course not offered in their own ethnic university were permitted to register elsewhere. By 1972 when the system was in full operation only 2 per cent of the 78,000 students attending residential universities in South Africa were registered at institutions for a designated population group other than their own. As in 1958, the largest such group was attending the University of Natal medical school (Figure 5.11). Academic staff were also race defined as only Whites could obtain posts at White universities, although the other universities were also staffed predominantly by Whites.

The ethnic exclusivity of the African universities was abolished in 1979 when, for example, the University of Zululand was permitted to admit African students other than those speaking Zulu or Swazi. Finally, in 1985 all universities were permitted to admit any student regardless of population group, marking one of the earliest breakdowns in the apartheid system.

The promotion of the homeland policy was also responsible for the creation of yet more universities. The newly independent states, Transkei, Bophuthatswana and Venda, all established new universities as one of the status symbols of separate statehood. Qwaqwa also set up a branch of the University of the North to which South Sotho-speaking students had previously been directed. Thus by the early 1980s there were separate residential universities for virtually every African linguistic group.

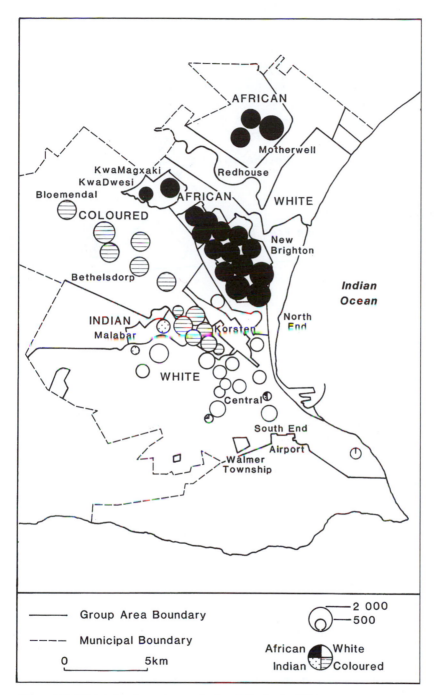

Figure 5.9 High school enrolments in Port Elizabeth, 1990
Source: Based on information supplied by the various education authorities and private schools

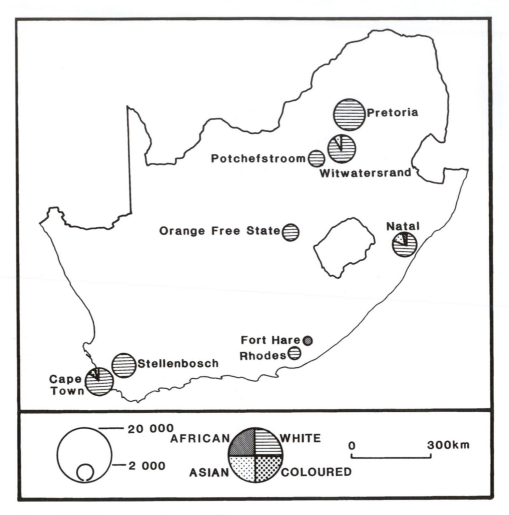

Figure 5.10 University enrolments, 1958
Source: Based on statistics in South Africa (1961) *State of the Union 1960–61*, Johannesburg:
Da Gama

More institutions were established to supplement existing facilities. In 1976 the Medical
University of South Africa was created outside Pretoria for the training of additional African
doctors. In 1983 Vista University was established to provide urban campuses for the African
population, where no branch of an African university existed in the suburbs under homeland
control. The aim of the institution was to provide tertiary education for African students in
the larger cities, outside the western Cape. As such, it marked a major departure from the
previous approach of sending students for further education back to the homelands in order
to foster national identities.

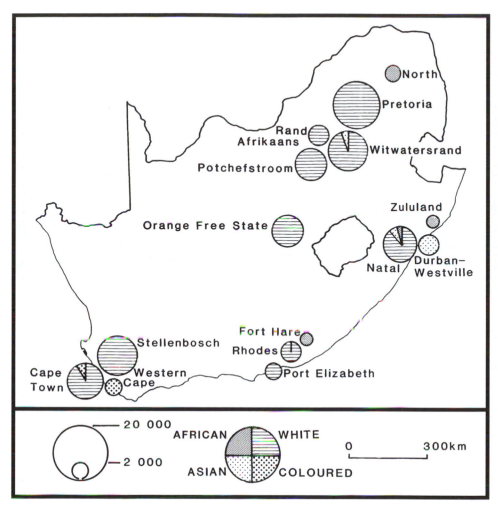

Figure 5.11 University enrolments, 1974
Source: Based on statistics in South Africa (1975) *State of South Africa 1974*, Johannesburg: Da Gama

PERSONAL SECURITY

It is remarkably difficult to evaluate spatially the various indicators of personal well-being, as most were not made available for small area analysis. However, some do lend themselves to geographical analysis when a statistical source is offered on a spatial basis. The differing levels of security illustrate aspects of living in a divided city through the availability of police station records, offered in response to questions raised by members of parliament.

The Cape Town metropolitan region has been taken as an example. In 1989 the region recorded 1,483 murders. This indicated the exceptionally high rate of approximately 70 murders per 100,000 population. The distribution of reported murders exhibited a marked concentration in the Coloured and African suburbs (Figure 5.12). This has several

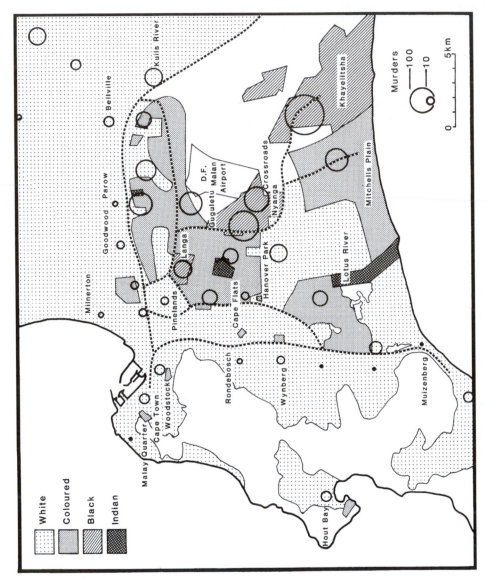

Figure 5.12 Murders in Cape Town, 1989
Source: Based on statistics in *Hansard*, 21 March 1990, cols. Q615–16

explanations related to the extent of socio-economic deprivation within the two communities, and to the concentration of policing in the White sections of the region. The particularly high number of murders in the Khayelitsha squatter area is indicative of this relationship. In contrast, the White suburbs on the western and northern sides of the metropolitan area recorded markedly fewer murders.

In some of the displaced communities on the Cape Flats the levels of gang warfare achieved frightening proportions as gangs struggled for territorial turf and control of protection racketeering and drug trafficking in some of the more notorious suburbs (Figure 5.13) (Pinnock 1984). The remarkably high murder rates among young Coloured males attracted considerable attention, with such headlines as 'even the Grim Reaper is not colour-blind' (*Weekly Mail*, 30 April 1992). In the age group 15–24, Coloured males were almost ten times as likely to be murdered as White males and twice as likely as African males of the same age group. The ongoing gang warfare gained greater publicity with the coming to prominence of the vigilante group, People Against Gangsterism And Drugs (PAGAD), in

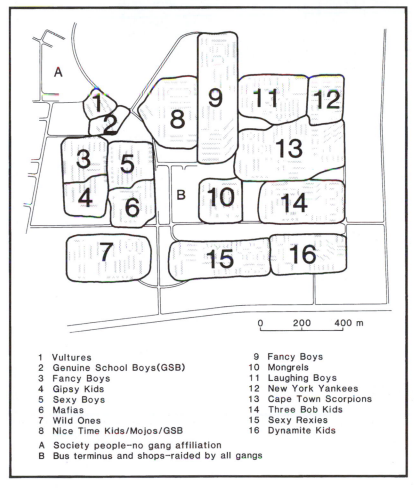

1	Vultures	9	Fancy Boys
2	Genuine School Boys(GSB)	10	Mongrels
3	Fancy Boys	11	Laughing Boys
4	Gipsy Kids	12	New York Yankees
5	Sexy Boys	13	Cape Town Scorpions
6	Mafias	14	Three Bob Kids
7	Wild Ones	15	Sexy Rexies
8	Nice Time Kids/Mojos/GSB	16	Dynamite Kids

A   Society people–no gang affiliation
B   Bus terminus and shops–raided by all gangs

Figure 5.13 Gang territories in Hanover Park
Source: Modified after D. Pinnock (1984) *The Brotherhoods: Street Gangs and State Control in Cape Town*, Cape Town: David Philip

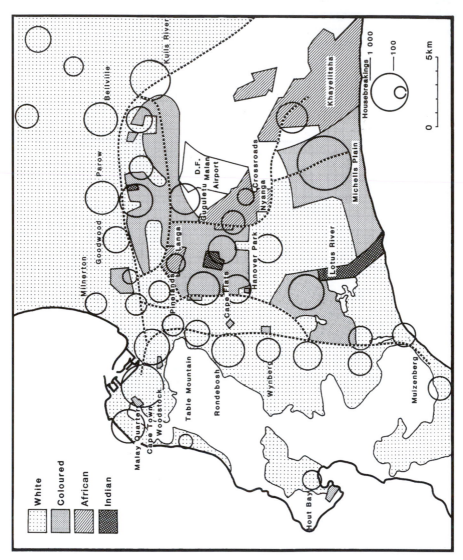

Figure 5.14 Housebreakings in Cape Town, 1989
Source: Based on statistics in *Hansard*, 21 March 1990, cols. Q615–16

1996. However, sections of the movement became involved in the wider conflicts of the Islamic world in the late 1990s and heightened the conflict in the Western Cape. This included the internationally high-profile bombing of the American-owned Planet Hollywood restaurant at the Cape Town Waterfront in 1998 (*Time*, 11 January 1999).

On the other hand, when reported housebreakings are examined the pattern was changed (Figure 5.14). Central Cape Town, which listed few murders, recorded the highest levels of housebreakings. Profitable suburban burglary in the White areas gave rise to intense activity by the security industry, and a change in the appearance of many streets as high walls with electronic security gates replaced low walls and hedges around houses. Furthermore, in the Coloured areas the comparatively affluent new suburb of Mitchell's Plain recorded the highest number of housebreakings. The African suburbs reported the lowest levels, suggesting both low levels of reporting to an unpopular police force and little economic incentive for the activity in a poverty-stricken community. Thus within the apartheid city the various communities had, as with so many aspects of life, markedly different security priorities and experiences of life.

6   Resistance to apartheid

In 1948 the National Party government embarked upon a series of radical new policies which encountered widespread opposition, notably from the African population which was most severely affected. The initial stage of opposition was the Programme of Action, when the policy of deputations and petitions was rejected in favour of boycotts, strikes and civil disobedience. However, opposition was not united and the Durban riots in January 1949, when 142 people were killed, illustrated the tensions between the Indian and African communities. In 1952 the joint Defiance Campaign was launched to obstruct the imposition of the multitude of repressive laws which were emanating from parliament. To counter opposition, the Suppression of Communism Act of 1950 was passed, with a remarkably flexible definition of what constituted the promotion of that ideology. Confrontation over the banning of the South African Communist Party led to more deaths, and to the virtual adoption of the party's programme and personnel by the African National Congress. This was symbolized by the adoption of the Freedom Charter in Kliptown in 1956, to the unease of the Africanists, who seceded to form the Pan Africanist Congress in 1959. Defiance campaigns and legal contests remained the main form of opposition throughout most of the 1950s, while the government systematically plugged the loopholes in the legislative and administrative structures of apartheid, and adopted further security legislation to enforce them.

THE BUILD-UP OF RESISTANCE

The introduction of new African local authorities legislation, agricultural betterment schemes and forced removals in the later 1950s led to physical confrontation in a number of regions, notably in the Transvaal in Sekhukuneland and the Hurutshe and Mamatlhola's Locations (Figure 6.1). In Pondoland in the Transkei in 1958–60 resistance to compulsory cattle dipping escalated into a mass revolt and the imposition of a state of emergency. The rising culminated in military action against organized rural rebels in the bloody engagement at Ngquza Hill and subsequent massive repression of anti-government activity.

The development of official policies, notably separate education, passes for African women, and the large-scale removals, resulted in general unease and resentment. When added to the continuous problems of poverty, squatter removals, liquor raids and bureaucratic intransigence, the potential for violence within African communities was high (Cobbett and Cohen 1988). In 1959 riots took place in Durban's African townships, and in the following year nine policemen were killed after a liquor raid in Cato Manor in Durban. Both the African National Congress and the Pan Africanist Congress organized

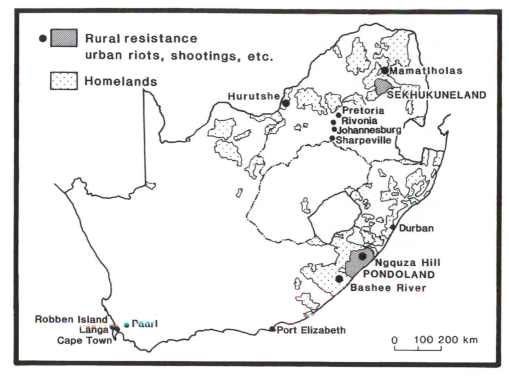

Figure 6.1 Early resistance to apartheid
Source: Compiled by the author

demonstrations in connection with a variety of grievances, more particularly the pass laws. On 21 March 1960 at a Pan Africanist Congress pass law demonstration at Sharpeville in the southern Transvaal, sixty-nine African people were killed by the police, initiating national and international protests. More protesters were killed at Langa in Cape Town on the same day.

The immediate government responses were the imposition of a state of emergency, and the banning of both the African National Congress and the Pan Africanist Congress. Both organizations thereupon formed military wings, Umkhonto we Sizwe (Spear of the Nation) and Poqo (Standing Alone) respectively, to attack installations and officials with the intention of seizing power by force. Riots in Paarl and the murders at Bashee River prompted a decisive response. The government was able within a couple of years to effectively crush such opposition, imprisoning the principal leaders of both organizations, and exiling the remainder. The arrest of the African National Congress leadership at Rivonia in 1963 proved to be decisive. The high-security prison on Robben Island off Cape Town accordingly held the leaders of all the major anti-government factions until their release in 1989–90.

ARMED RESISTANCE

The exiled organizations established links with a number of countries for the training of their personnel and launching of diplomatic initiatives (Figure 6.2). Although the Africanist

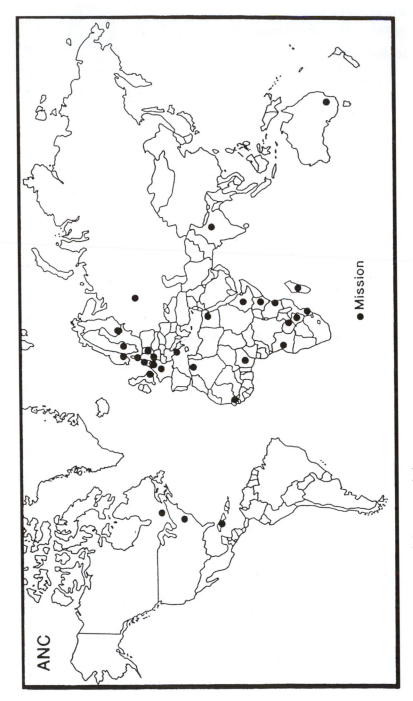

• Mission

ANC

Figure 6.2 African National Congress missions
Source: Based on information in S. Thomas (1996) *The Diplomacy of Liberation: The Foreign Relations of the African National Congress since 1960*, London: Tauris Academic Studies

ideals of the Pan Africanist Congress met with support in many newly independent African countries, the majority were far removed from the series of anti-colonial wars and insurgencies which were fought in southern Africa from the mid-1960s onwards. In contrast the African National Congress, with its adherence to non-racialism, was more restricted in its active international support base, but included countries and movements closer to South Africa (Thomas 1996). Indeed, it is notable that the liberation organizations maintained a wider network of quasi-diplomatic missions than the government in Pretoria.

Camps were established in Tanzania and later in Angola, Mozambique and Zambia for the instruction of the military wings of both the African National Congress and the Pan Africanist Congress (Figure 6.3) (Davis 1987). Specialized training was undertaken in a number of sympathetic countries, notably in the Soviet Union and its allies and more distant African states such as Libya and Uganda. The headquarters of the African National Congress were established in Dar es Salaam, but moved to Lusaka after the Morogoro conference in 1969,

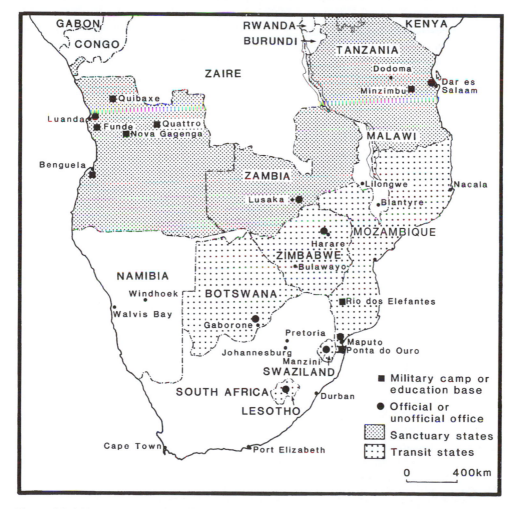

Figure 6.3 African National Congress bases
Source: Modified after S.M. Davis (1987) *Apartheid's Rebels*, New Haven, CT: Yale University Press

in order to be closer to South Africa. At that conference it was recognized that armed struggle was the only way to achieve the aims of the organization.

After 1976 the African National Congress was faced with the task of sustaining a substantial number of refugees from South Africa (Mbeki 1996; Meli 1988). The number of exiles increased from possibly 1,000 in 1975 to 9,000 five years later. In order to accommodate this increase a new centre was established at Mazimbu, near Dar es Salaam, in 1979. Education and training were provided at the Solomon Mahlangu Freedom College. The period after 1975 was one of reorganization as bases were established closer to South Africa in an attempt to intensify armed resistance. Angola became the main African National Congress military centre with bases in areas under the control of the Angolan central government. The attempt to establish bases in Mozambique was temporary as the 1984 Nkomati Accord between the South African and Mozambican governments led to their removal. Similarly, activity in Zimbabwe was ended with the election of the Zimbabwe African National Union–Patriotic Front government in 1980, which backed the Pan Africanist Congress. Thus the African National Congress activity was largely confined to the three sanctuary states of Angola, Zambia and Tanzania. Its resources were spread widely in order to prevent undue influence being exerted by any one state.

SOWETO RIOTS, 1976

By 1976 the long period of apparent security force supremacy had begun to wane. The independence of Angola and Mozambique under Marxist governments, and the increasing South African military involvement in the ensuing civil wars in both, brought the prospects of African empowerment closer. The authorities on the other hand pursued their policies with increasing insensitivity and triumphalism, thereby breeding intense resentment. The point of conflict came over the imposition of new regulations attached to the Bantu Education programme, which sought to introduce the compulsory use of Afrikaans as the medium of instruction for mathematics, geography, physical science and biology in secondary schools, 'for the sake of uniformity' (Bonner and Segal 1998: 81). The burden of this requirement, added to the perceived inferiority of the programme itself, resulted in protests and demonstrations, organized by the South African Students' Movement.

The date of 16 June 1976 was chosen for country-wide demonstrations against the imposition of Afrikaans in African schools. The involvement of many strands of discontent allowed the whole situation to lead to riots. A march by school children in Soweto ended in confrontation with the police as the necessary permission from the West Rand Bantu Affairs Administration Board had not been given. The police opened fire, killing a number of young demonstrators. Serious riots broke out the following day with associated murder, arson and looting. The particular targets of the rioters were government properties, notably those administered by the Bantu Affairs Administration Boards, including libraries, clinics and most significantly beerhalls, which provided the revenues for the system. Schools, shops and houses were also attacked (South Africa 1980).

The riots spread from Soweto to other towns on the Witwatersrand as well as to Pretoria and its vicinity (Figure 6.4). Extensive damage was done to the Universities of the North and Zululand. It was only in August that significant rioting began in the western Cape, when it had largely died down on the Witwatersrand. The African areas in Cape Town were affected in mid-August and the Coloured areas in early September. Riots in Port Elizabeth in mid-August were similarly directed against Administration Board property. Sporadic

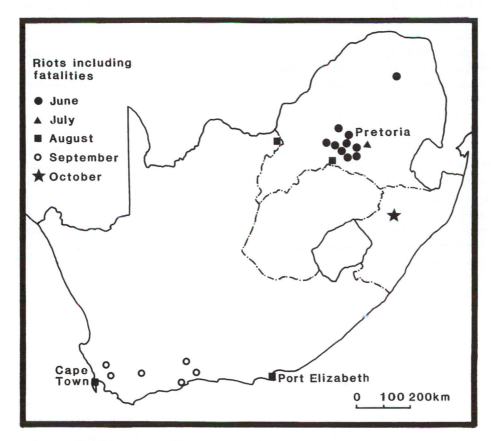

Figure 6.4 The diffusion of violence, 1976
Source: Based on information in South Africa (1980) *Report of the Inquiry into the Riots in Soweto and Elsewhere*, Pretoria: Government Printer

outbreaks occurred elsewhere in the country but were not as sustained as those on the Witwatersrand and Cape Town. Almost half the 575 deaths occurred in Soweto alone. Natal and the central areas of the country experienced little serious threat to state control. The government was able to restore its authority, and embarked upon a limited reform policy. For several thousand young Africans, however, the riots provided the opportunity to flee the country and join the various exiled organizations seeking the overthrow of the South African government. The role of pupils and students had been of major significance in the origin and continuation of the riots and the breakdown of authority, whether state or parental.

1984 AND THE 1985–90 STATES OF EMERGENCY

The appointment of P.W. Botha as Prime Minister in 1978 led to the implementation of a programme of controlled reform of the apartheid system. At the same time the state security structures were elaborated and made more effective. The opposition also regrouped, with the creation of such organizations as the Soweto Civic Association and the Port Elizabeth

Black Civic Organization in 1979. An umbrella organization, the United Democratic Front, was established in 1983 to co-ordinate the activities of the extra-parliamentary opposition. The introduction of the new tri-cameral constitution in 1984, with the structural exclusion of the African population from the decision-making process of the country, led to a general state of unrest ending the comparative quiescence of the early 1980s.

In September 1984 riots broke out in a number of African townships in the southern Transvaal, notably in the Vaal Triangle. The death rate due to political violence increased sharply. In addition to the new constitution, rent and transport fare increases, together with a school boycott, created an explosive environment. Discontent with the new structures of African local government, which were held responsible for many of the townships' ills, also provided a visible target for the rioters. Meetings were banned and the army was widely deployed in support of the police in controlling demonstrations, searching houses for arms and maintaining order. In the absence of any legal outlet, funerals became the main occasion for political gatherings.

Boycotts and demonstrations spread to other parts of the country in early 1985, notably to the eastern Cape. Furthermore, much of the unrest was experienced in the smaller towns, such as Cradock, Colesberg and Graaff-Reinet. On 21 March 1985, the twenty-fifth anniversary of the Sharpeville shootings, police opened fire and killed twenty mourners in a funeral procession in Langa Township, Uitenhage. The incident was highly inflammatory, both internally and internationally. Violence subsequently became particularly deadly in a number of centres on the East Rand.

A State of Emergency was declared in July 1985 in several eastern Cape and East Rand districts (Figure 6.5) (Lemon 1987). Wide powers of arrest and detention were given to the police. Major drives against anyone suspected of violence or incitement ensued, although the majority so detained were released quite rapidly, owing to the exceptionally large numbers involved. Press censorship was imposed, thereby hiding the widespread activities of the security police and their death squads. The emergency was lifted in a number of districts in October. However, violence spread to the western Cape and a state of emergency was imposed there in the same month. The emergency was lifted in more districts in December, and in its entirety in March 1986.

The 1984–6 period was preceded by raised expectations on the part of the African population as the reform process appeared to offer prospects of change, notably in the liberalization of the economy. However, Africans were excluded from the 1984 constitution which provided only for White, Coloured and Indian houses of parliament. Furthermore, the economy was failing to grow at a rate to provide jobs for school leavers or any material improvement in living conditions. The African political parties and trade unions, newly legalized in 1979, thus had ample ground upon which to work to gain support. Indeed, in 1985 the Congress of South African Trade Unions was formed which gave a significant boost to the power of organized labour.

Levels of violence continued to be high and a further state of emergency was imposed on the entire country in June 1986 (Figure 6.6). A dramatic lessening of violence and politically motivated murder ensued, as massive detentions were carried out and the organizers of the various factions were scattered. Relative order returned to the country, if under a considerable degree of repression. Despite the appearance of peacefulness in 1987, the initiation of civil strife in Natal and KwaZulu in the second quarter marked a new phase of conflict (Figure 6.7). By 1989 the scale of fighting and political murder in Natal exceeded that under the 1986 state of emergency for the country as a whole. The struggle between the Inkatha Freedom Party and the African National Congress-affiliated groups was such that the

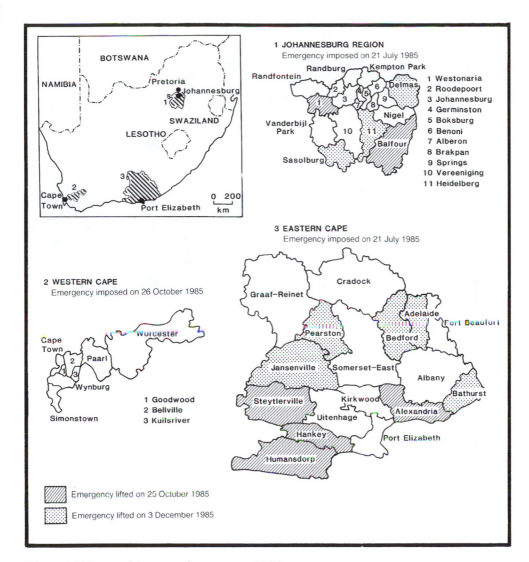

Figure 6.5 Extent of the state of emergency, 1985
Source: After A. Lemon (1987) *Apartheid in Transition*, Aldershot: Gower

resources of the KwaZulu government and indeed other, then secret, sources of support had to be invoked to prevent the collapse of the Inkatha movement (Adam and Moodley 1992). Little headway was made by the police or security forces in containing and preventing the violence as police death squads added to the levels of mayhem.

DEATHS IN DETENTION

As a part of the State of Emergency declared in 1960 following the Sharpeville killings, the government introduced detention without trial. In 1963 under the General Laws

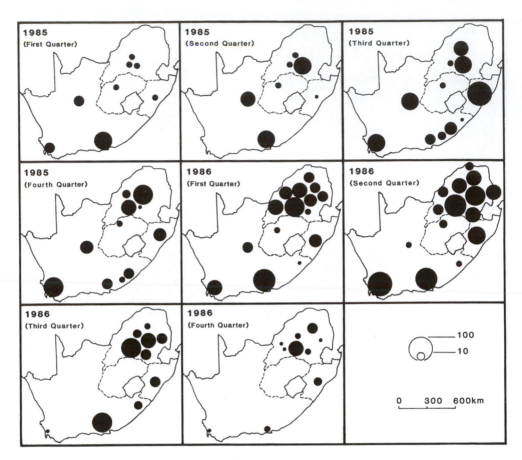

Figure 6.6 Deaths due to political violence, 1985–6
Source: Based on information supplied by the South African Institute of Race Relations

Amendment Act, later expanded as the Internal Security Act, this power became a permanent part of the state apparatus in fighting opposition. The period of detention was lengthened, and then made indefinite if sanctioned by a judge in 1966. Even such cursory authorization became unnecessary after 1976. The Transkei, Bophuthatswana, Venda and Ciskei governments introduced their own, often more draconian, legislation on attaining independence.

In the course of the period from 1960 to 1990 some 78,000 people were detained without trial (Coleman 1998). One of the notable features of this legislation was the virtually unlimited power exerted by the Security Police over the detainees, and the consequent number of deaths in detention, often after torture (Figure 6.8). In the 1960s two or three detainees died each year, until the widely publicized death of Ahmed Timol in 1971. No more deaths were recorded until the Soweto riots of 1976. There then followed a horrific twenty-six deaths in two years. The international outcry over the much publicized death of Black Consciousness leader Steve Biko in 1977 again brought a halt to the deaths. However, in the 1980s numbers of deaths once more began to rise until the end of the final State of Emergency. Thus some seventy-three people died in detention in the period, approximately

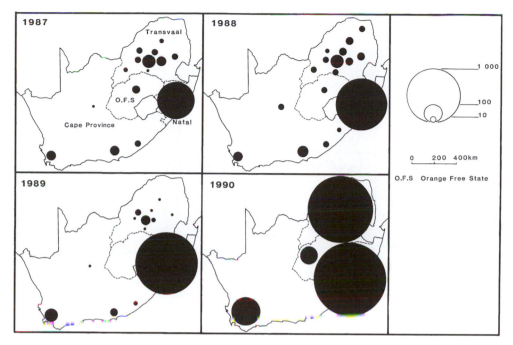

Figure 6.7 Deaths due to political violence, 1987–90
Source: Based on information supplied by the South African Institute of Race Relations

one person per 1,000 held. The figures do not include those who died in police custody but not specifically related to the enforcement of security laws. The ruthlessness of the Security Police was encouraged by the ethos of the government from senior ministers downwards. Minister of Police J. Kruger's comment on the death of Steve Biko – 'I'm not pleased, nor am I sorry; Biko's death leaves me cold' – is eloquent testimony to the official lack of concern at the methods employed (*Time*, 26 September 1977).

The Security Police operated widely not only from their own regional headquarters, but also from local police stations in the smaller towns. Consequently although the major cities, such as Johannesburg (fourteen deaths) and Pretoria (ten deaths), dominate the list, the smaller centres and homelands, particularly the Transkei, feature prominently. Individual buildings, notably John Vorster Square (now Johannesburg Central) police station (seven deaths), Pretoria Prison (five deaths) and Johannesburg Fort and the Sanlam Building (now Steve Biko House) in Port Elizabeth (both four deaths), feature prominently in the Security Police's drive against opposition. The southern Transvaal and the eastern Cape emerge as the most dangerous regions for political opponents of the government.

MILITARIZATION

Under the rule of Prime Minister, later President, P.W. Botha (1978–89) the country underwent a period of militarization (Cock and Nathan 1989). The South African Defence Force was expanded to meet the perceived 'total onslaught' of international communism

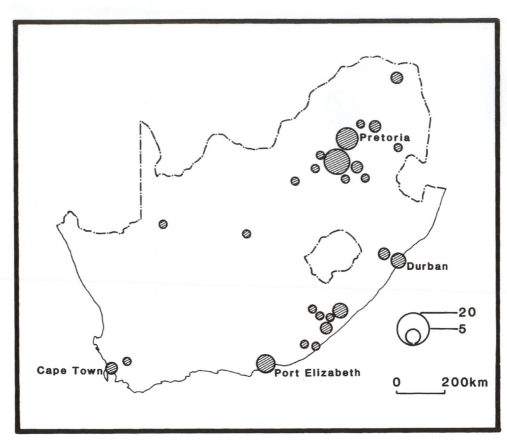

Figure 6.8 Deaths in detention, 1963–90
Source: Based on information supplied by the Human Rights Commission

(see Chapter 7). In the process new bases and new facilities were established for both the army and the police force (Figure 6.9). Some 600,000 hectares were appropriated to serve the needs of the armed forces (South Africa 1998a). Some of the extensive training terrains, including the P.W. Botha Battle School at Lohatla in the Northern Cape, were acquired as a result of the expropriation of African-occupied lands in the course of the homeland policy. Lohatla, extending over some 145,000 hectares, was one of the most extensive training facilities in the world outside the United States and the former Soviet Union. The army base complex at Voortrekkerhoogte at Pretoria was expanded to provide the key centre with air support, specialized hospitals, training centres, experimental laboratories (including those for chemical warfare), etc. In addition, the naval base at Simonstown was supplemented by the adjacent naval monitoring facilities at Silvermine. Weapons ranges were established in the Kalahari and on the coast as the South African army developed its rocket-firing capabilities (and secret nuclear weapons capability).

It was against the state's military facilities that the African National Congress scored its most publicized operations. The Umkhonto we Sizwe attack on Voortrekkerhoogte in 1981 was a highly significant symbolic act of defiance, demonstrating the organization's ability to reach the centre of the state's security apparatus. The car bomb explosion outside the South

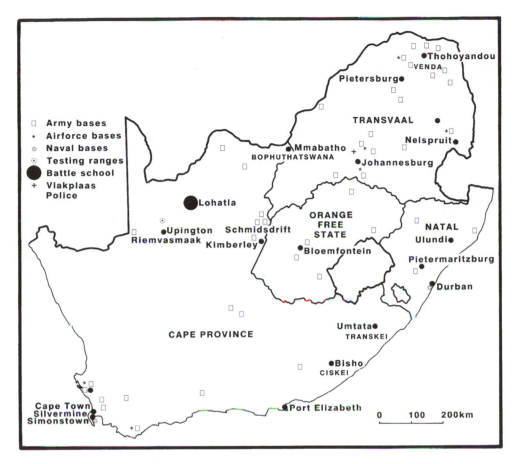

Figure 6.9 Military and police sites
Source: Modified after South Africa (1998) *Defence in a Democracy: South African Defence Review 1998*, Pretoria: Government Printer

African Air Force headquarters in Church Street in central Pretoria in 1983, which killed nineteen people and sparked another wave of retaliatory attacks, constituted the most spectacular operation undertaken by the organization. Economic sabotage through attacks on the strategic oil plants Sasol and Secunda in 1980 and the Koeberg nuclear plant in 1982 were again largely symbolic, demonstrating the vulnerability of the economy.

Parallel with the large military installations were the secret police, Special Branch, facilities, designed to supplement the regular structures. The training and command headquarters on the farm Vlakplaas outside Pretoria was established in 1978 as the key centre. Although only 99 hectares in extent, Vlakplaas acted as the base for the hit squads which carried out an officially sanctioned political assassination campaign against perceived opponents of the government in the 1980s and 1990s (Coetzee 1994). It also acted as a trans-shipment centre for the distribution of arms and ammunition to other sympathetic militant organizations including the Inkatha Freedom Party prior to the 1994 election.

The counter-revolutionary administrative structures erected by the government, under the secretive National Security Management System in 1979, were controlled by the State

Security Council (initially established in 1972) chaired by the President. A parallel set of structures reporting to the Council were the regional Joint Management Centres, ultimately based in each of the planning regions, and sub-centres for the Regional Services Councils and some 354 mini-management centres in individual municipalities and local areas (Alden 1996). The object of these structures was to integrate leaders of the administrative and business communities into the security apparatus. Co-ordination was aimed at the more effective implementation of the official strategies to counter and ultimately eliminate members of the internal opposition, or euphemistically, 'permanently remove from society' (*Mail & Guardian*, 28 May–3 June 1999: 3).

7 International response to apartheid

South Africa gained its dubious prominence in international politics in the post-Second World War era, through the almost universal condemnation of the policy of apartheid as violating internationally acceptable norms (Klotz 1995). However, several other related issues further complicated the international relations of the state. Prominent among them were the question of the South African administration of South West Africa (Nambia), which bedevilled relations between the United Nations and South Africa until 1990, and the expansionary intent of South African governments until the 1960s with regard to the then British High Commission territories (now Botswana, Lesotho and Swaziland). Further issues, notably economic and political sanctions imposed by other states against South Africa and its retaliatory destabilization of its neighbours, led to a significant and possibly decisive impact of external affairs on the pursuit of apartheid.

In the regional context South Africa operated as the local superpower, exerting direct influence upon a wide range of countries in the southern part of Africa through its economic strength (Figure 7.1). The Southern African Customs Union and the Rand Currency Area, the former dating from the time of the formation of the Union in 1910, were important organizations for exerting economic influence. Furthermore, the configuration of the transportation systems, notably the railway network built in colonial times, focused international trade routes upon the harbours of South Africa. This position was exploited by the South African government in its attempts to obtain international recognition of its apartheid policies.

DIPLOMATIC SANCTIONS

South Africa was subjected to intensive diplomatic pressures by countries opposed to the principle of institutionalized racial discrimination. This was evident not only in the volume of hostile resolutions adopted by the United Nations, the Organization of African Unity and many other international bodies, but in the country's exclusion from participation in their workings. In addition, there was a remarkably low level of foreign diplomatic representation within the country, and a limited range of South African official accreditation in other countries. The South African government thus maintained a diplomatic network far smaller than its size and power would have suggested.

As a result of the condemnation of apartheid policies, the country was relegated to the status of an international pariah (Figure 7.2). This was a gradual process rather than a sudden occurrence, and reflected the propaganda successes of the anti-apartheid movement allied to the African National Congress and the Pan Africanist Congress. Thus South Africa

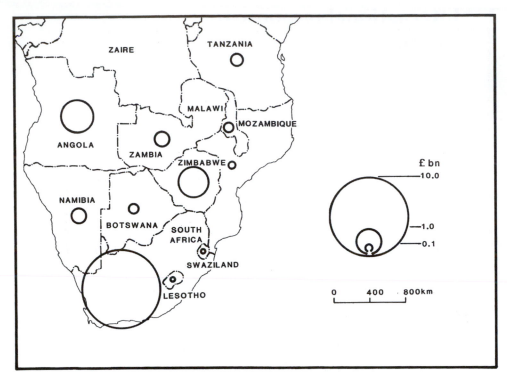

Figure 7.1 Southern Africa: government expenditure, 1985–6
Source: Based on statistics in *Statesman's Yearbook*, 1985–6

maintained some twenty-one diplomatic missions abroad in 1955 and twenty-five in 1989, although the number of independent states worldwide had increased markedly in the meantime. The diplomatic missions maintained in Transkei, Ciskei, Bophuthatswana and Venda are excluded from this discussion. These four states remained totally isolated internationally except for mutual and South African recognition for the entire period of their existence.

In the late 1960s and early 1970s the South African government attempted to break out of this isolation by extending its network, mainly with military and fascist-orientated regimes in Latin America. However, the gains were ephemeral as increasing pressures were exerted to isolate the country. The distribution of government missions was heavily concentrated in western Europe, reflecting long-term links. Most areas of the world were effectively closed to South African diplomats, and indeed to South African passport holders. Of significance was the lack of official relations with other African states, with the exception of Malawi. In Asia diplomatic relations were only maintained with the two other international outcasts, Israel and the Republic of China (Taiwan) (Geldenhuys 1990). Elsewhere only the United States, Canada and a small group of Latin American states retained formal links.

In a number of cases inter-governmental lines of communication were maintained at a lower level, such as through a consul or trade representative, most notably with Japan. In certain instances these were downgraded from the previous full diplomatic representation. In others they formed a new link which, owing to the general political climate, could not be at embassy or legation level. Thus the consulate in Maputo ceased to function after the

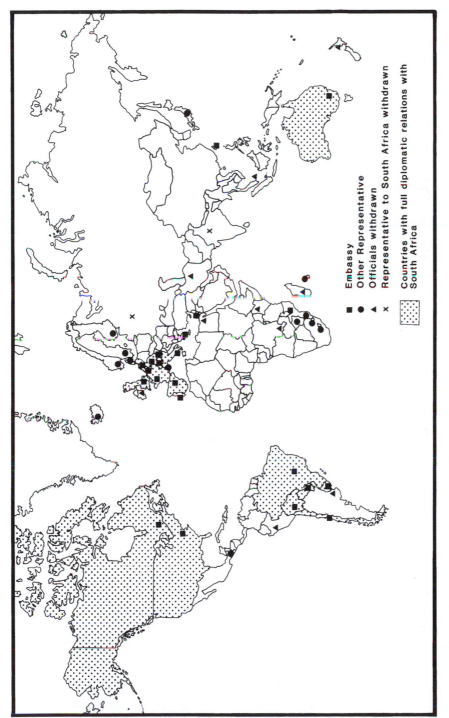

Figure 7.2 South African diplomatic representation, 1989
Source: Based on information in various issues of South Africa (1950–90) *Official Yearbook*, Pretoria: Government Printer

Portuguese withdrawal in 1975, but was reopened in 1984 following the Nkomati Accord, although full diplomatic relations were not established. Similarly after Zimbabwean independence in 1980 South Africa maintained a trade representative in Harare.

The low level of foreign contacts was partially self-imposed, as symbolized by the country's withdrawal from the Commonwealth in 1961. Relations with countries such as the Soviet Union were terminated at South Africa's insistence as part of its racial and anti-communist policies. India terminated its representation in protest at overt racial discrimination and the political deterioration of the position of the South African Indian community. However, the main cause of the loss of representation resulted from sanctions, mostly imposed at independence in the case of African states. The diplomatic efforts to solve the South West African independence issue, initiated with the protocol signed in Brazzaville in 1989, began the process of South Africa's re-entry into the mainstream of diplomatic representation.

AIRWAYS SANCTIONS

One of the most spectacular spatial examples of the imposition of sanctions was that applied to air communication with South Africa (Griffiths 1989; Pirie 1990b). In 1962 the United Nations called upon member states to refuse landing and overflying rights to South African aircraft. The reciprocal refusal to allow other airlines to fly to South Africa was taken subsequently by a number of countries as a logical extension of sanctions.

The main destination of travellers from South Africa was western Europe. Thus in 1963 a number of African states were able to take advantage of their strategic position and refused to allow South African Airways the use of their airspace (Figure 7.3). This necessitated the re-routing of flights to Europe around the west coast of Africa, instead of the previous more direct route via Nairobi. South African Airways was forced to absorb the increased costs of the longer route (approximately an extra 1,400 kilometres) and, by using the then Portuguese airports at Luanda and at Ilha do Sal in the Cape Verde Islands, to overcome the problems of restricted aircraft range. After independence Angola joined the boycott, but the Cape Verdean government was too dependent economically on the landing fees to do so. The introduction of longer-ranged aircraft enabled South African Airways to reduce its dependence upon intermediate airports except in emergencies. Thus the later acquisition of landing rights at Abidjan in the Ivory Coast was the result of political, not operational, policy. The extreme consequence of the ban on overflying was the routing of South African flights to Tel Aviv around the west coast of Africa, thereby doubling the flying time and the distance covered.

Other international routings were more problematical, reflecting the country's external relations. The direct air link with the United States proved to be politically the most contentious. Inaugurated in 1969, the flights were the subject of controversy from the beginning. However, it was only in 1986 with the major thrust of the Comprehensive Anti-Apartheid Act that bans were placed on South African Airways and on American airlines from operating between the two countries.

Links across the Indian Ocean were equally problematical and politicized. The Australian airline, Qantas, began operating between Sydney and Johannesburg as part of the Commonwealth link-up programme in 1952, and South African Airways began a reciprocal service five years later. However, the political pressures of the post-1984 era were such that in 1986 the Australian government prevented further direct flights by both airlines between

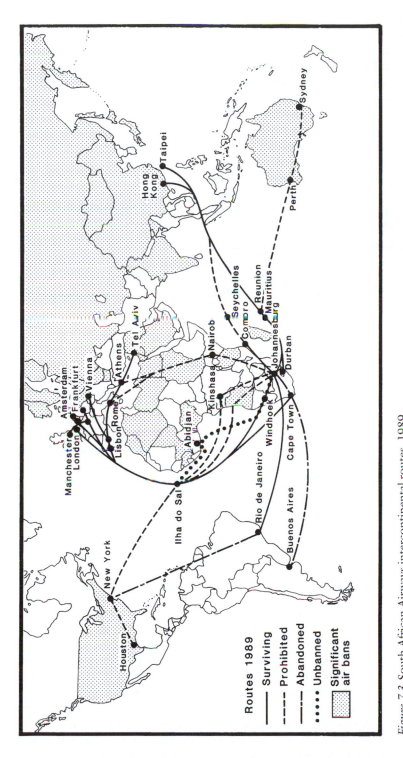

Figure 7.3 South African Airways intercontinental routes, 1989
Source: After G.H. Pirie (1990) 'Aviation, apartheid and sanctions: air transport to and from South Africa 1945–1989', *GeoJournal* 22: 231–40

the two countries. Qantas then flew to Harare, necessitating inconvenient stopovers for South African passengers, while South African Airways flights to Hong Kong provided an alternative routing.

South African Airways established routes to Hong Kong and Taipei, via Mauritius, but direct links with Japan and India eluded the corporation. In South America, links with Brazil and Argentina were established, but the latter were withdrawn as a result of unprofitability, although the collapse of the military dictatorship in Argentina undoubtedly contributed to the decision, as symbolized by the downgrading of diplomatic relations.

Within the southern African region, South African Airways links with Zimbabwe were retained at independence, as were those with Malawi, Zambia, Mozambique, Botswana, Lesotho and Swaziland. Within the Indian Ocean, Mauritius retained links, while the Seychelles banned flights in 1980. Flights to Moroni in the Comoros were commenced following the development of a Southern Sun Hotels holiday resort on the island. Homeland airlines, including Transkei Airways and Ciskei International Airlines, failed to establish any links outside South Africa.

Sanctions were thus effective in reducing the overseas range of South African Airways operations. However, they continued, and in many cases the increased numbers of flights offered by a wide range of European airlines effectively enabled links to be maintained with the rest of the world.

THE HIGH COMMISSION TERRITORIES

Until South Africa withdrew from the Commonwealth in 1961, one of the constant aims of successive governments had been the incorporation of the three British protectorates of Basutoland, Bechuanaland and Swaziland (Hyam 1972). Provision had been made in the British South Africa Act of 1909, which established the Union of South Africa, for the eventual inclusion of the three territories, as well as Southern Rhodesia (Zimbabwe) (Figure 7.4). The electorate of the latter territory in a referendum in 1923 decided not to join the Union. Various schemes of incorporation of the three remaining territories were negotiated in the period up to 1948. However, they all failed to materialize as successive British governments, of whatever persuasion, remained suspicious of South African intentions towards the indigenous populations and the preservation of their land rights.

The National Party government after 1948 remained equally committed to incorporation. Indeed official reports, including that of the Tomlinson Commission, drew maps and published statistics effectively including the three High Commission territories in the African areas of South Africa (see Figure 3.2, p. 71). Accordingly, it was possible to suggest that the indigenous population had retained half the land area of 'greater' South Africa. The three protectorates thus were initially viewed as part of the grand apartheid design.

Economically they operated as integral parts of the country through the currency and customs union and, until 1963, the absence of most forms of control over the borders between the states. Although South Africa failed to gain sovereignty over the countries, it was able to exert influence over them, as witnessed by the successful pressure applied over the marriage of Sir Seretse Khama, leading to his removal from the chieftainship. However, once South Africa left the Commonwealth the possibility of political incorporation was not considered again and the three states became independent in the later 1960s.

In 1954 the British government launched the Federation of Rhodesia and Nyasaland on the political platform of partnership between Europeans and Africans. In this respect it was

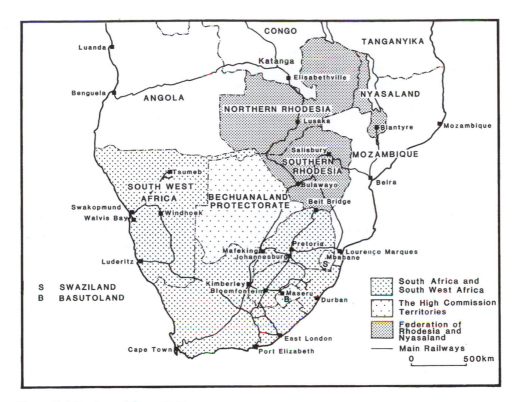

Figure 7.4 Southern Africa, 1961
Source: Compiled by the author from various sources

designed to counter South African influence in the region and as a demonstration that there was a better alternative to apartheid. The new federation lessened its economic dependence upon South Africa through such schemes as the direct rail link to Lourenço Marques and the promotion of manufacturing industry. However, rising African nationalism led to its disbandment in 1963 and the independence of Zambia and Malawi the following year. The future of Southern Rhodesia, following its unilateral declaration of independence (UDI) from Great Britain in 1965, remained a matter of confrontation for a further fifteen years. South Africa was drawn into the issues of international sanctions and armed conflict against Rhodesia in an attempt to keep insurgents away from its own borders. Limited police support was offered, but military involvement was avoided, as was the imposition of sanctions. Eventually South Africa was instrumental in securing the country's transition to African government, when continued White control appeared to be unsustainable in 1979–80.

SOUTH WEST AFRICA

The legal relationship between South Africa and South West Africa was one of the most significant international problems to confront successive South African governments between 1915 and 1990 (Wellington 1967). In 1915 South African forces occupied the

German colony. In 1919 the League of Nations awarded South Africa a mandate to administer the territory on its behalf. The mandate was held under class C, which placed few
restrictions upon the freedom of the mandatory power to rule as it thought fit, as no
concepts of self-government were written into this category of trusteeship. The South African government was allowed to administer the territory as an integral part of the Union,
limited only by an obligation 'to promote to the utmost the material and moral well-being
and social progress' of its indigenous inhabitants (South Africa 1964: 47). In the main they
were confined to a series of reserves (Figure 7.5). However, with the benefit of proximity the
government was able to encourage South African, mainly Afrikaner, land colonization and
the extension of White control on the same lines as in South Africa.

    After the Second World War the League of Nations mandated territories were transferred
to the trusteeship of the United Nations. The South African government refused to make
the transfer because greater obligations, notably those of indigenous political advancement,
were imposed. Instead, the government sought to incorporate the territory into South
Africa, but this course of action was rejected by the United Nations in 1946. The status of
the territory thus became a matter of bitter dispute, with the United Nations attempting to
exert influence over the activities of the mandatory power, while South Africa fended off
what it considered to be interference in its internal affairs. After 1948 the government
refused to render any more reports in terms of the mandate and the following year it took a
step closer to annexation by providing for the representation of the territory's White population in the South African parliament. In 1966, after lengthy proceedings at the International
Court of Justice at The Hague, the United Nations revoked South Africa's mandate. During
the following year a Council for South West Africa was established, under a Commissioner
charged with the task of preparing the territory for independence. In 1968 the international
body symbolically changed the name of the territory to Namibia. The United Nations thus
gained a direct interest in the internal affairs of Namibia, where theoretically it did not have
one in the internal affairs of South Africa. This made the international campaign against
apartheid easier, as apartheid was applied to what was legally an international territory.

SEPARATE DEVELOPMENT IN SOUTH WEST AFRICA

The National Party government began to apply South African apartheid laws to South West
Africa after 1948. Influx controls were imposed and passes were demanded of African
migrants to the towns. In 1964 the government published the Odendaal Report on the
affairs of South West Africa (South Africa 1964). Following its recommendations in 1968
the government proposed the establishment of some ten separate administrations for the
various ethnically defined indigenous population groups, along the same lines as the system
introduced to South Africa (Figure 7.6). The ten homelands covered some 326,000 square
kilometres, or nearly 40 per cent of the area of the territory. In several cases the numbers of
people involved were too small for an effective government to be brought into being.
Accordingly, the representatives of the approximately 30,000 Bushman (San) community
consistently sought to remain under the control of the central government, while the fewer
than 10,000 Tswanas were able to support only limited government functions.

    The initial homeland was the Rehoboth Gebeid which had enjoyed a measure of self-
government dating back to German colonial times. This was followed by Ovamboland in
1967. The Ovambo population constituted approximately half the total of the country, and
formed the numerically and politically dominant group, in marked contrast to South Africa

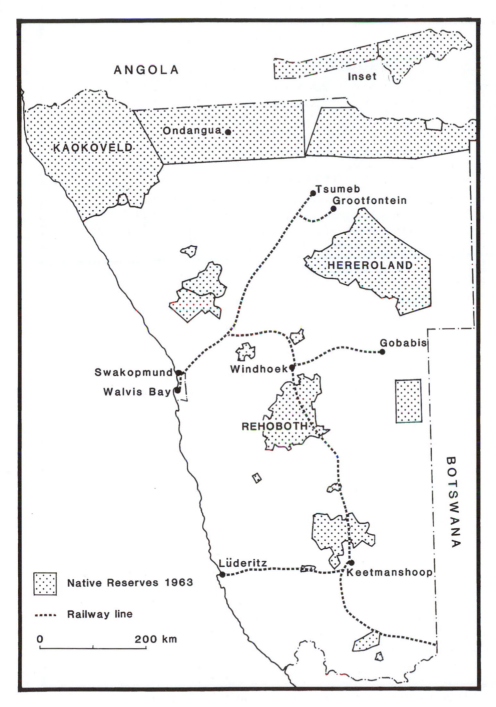

Figure 7.5 South West Africa Native Reserves
Source: After South Africa (1964) *Report of the Commission of Enquiry into South West African Affairs 1962–63,* Pretoria: Government Printer

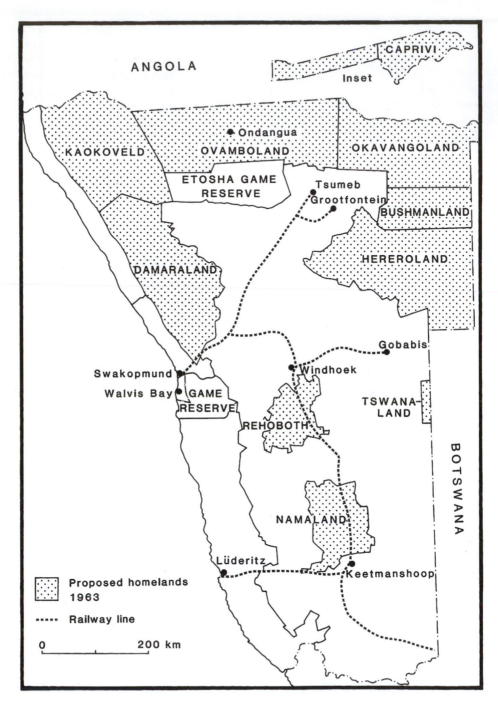

Figure 7.6 South West Africa homelands
Source: After South Africa (1964) *Report of the Commission of Enquiry into South West African Affairs 1962–63*, Pretoria: Government Printer

where no African group was so preponderant. Separate governments for Okavangoland, East Caprivi and Kaokoland followed in the early 1970s. Thereafter the administrative process continued with or without the support of elected bodies.

The development of the homeland policy, however, was entangled with the international situation. The concept of a unitary state was reinforced as the United Nations and South Africa renewed contact, through the offices of a group of Western powers. In 1975 the government established the Constitutional Conference on South West Africa (Turnhalle Conference) based on representation by ethnic group. The homelands were thus reduced to a second tier of government, not potential independent states. In 1978 a country-wide election was held for a Constituent Assembly, which was reconstituted as the National Assembly the following year. In 1980 ethnic elections were held for the homeland administrations, except in Bushmanland, whose leaders could not support the government apparatus, the Rehoboth Gebeid, where the election had taken place in 1979, and most significantly Ovamboland, where the security situation was such that the holding of reasonably free elections was impossible.

In the urban areas, much of the South African legislation applicable to Africans was introduced under the Natives (Urban Areas) Proclamation of 1951. However, the Group Areas Act was considered to be too sensitive and so it was not replicated in the territory (Simon 1986). Instead reliance was placed upon restrictive covenants in title deeds and the manipulation of the construction and allocation of state housing to separate the White and Coloured populations. The towns in South West Africa were already highly segregated, and new, post-1950, planning was carried out within the apartheid framework.[1] In Windhoek, the Old Location, adjacent to the city centre, was demolished in the 1960s and replanned as a White suburb, while the African population was moved 5 kilometres north to Katutura, which in Herero inappropriately could be translated as 'the place where we do not settle' (Herbstein and Evenson 1989: 7). A segregated suburb, Khomasdal, was set aside for the Coloured population in addition to the already segregated African population (Figure 7.7). In view of the fact that at the time of the 1970 census the White population constituted only 40 per cent of the city's total, the disparity in land allocation was marked.

The complex international negotiations which followed in the course of the 1980s duly led to the independent state of Namibia in 1990, following elections supervised by the United Nations Transition Assistance Group. During this period, although the second tier of administration was based on ethnic lines, much of the racially discriminatory legislation was repealed in 1977. The Abolition of Racial Discrimination (Urban Residential Areas and Public Amenities) Act of 1979 removed all residential restrictions within the towns and the following year the segregation of amenities was abolished. However, many social services, including education, were provided at ethnic authority level, thus retaining many aspects of apartheid until independence.

At independence there remained a number of unresolved territorial problems between South Africa and Namibia, dating from German colonial days (Berat 1990). These included the ownership of the Walvis Bay exclave and the Guano islands along the Namibian coastline and the definition of the Orange River boundary (Simon 1995). These were only resolved, to Namibia's satisfaction, just before the 1994 elections in South Africa. In 1998 and 1999 the apartheid legacy returned briefly with the unsuccessful attempt by East Caprivi to secede from Namibia.

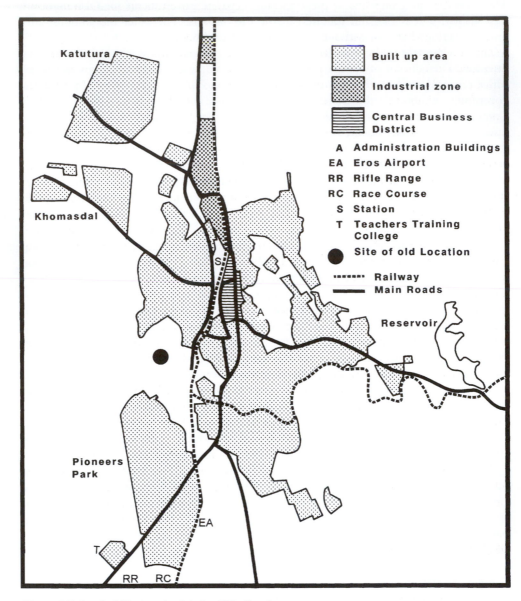

Figure 7.7 South Africa's colonial city: Windhoek
Source: Modified after 1:50,000 official topo-cadastral map

INSURGENCY

The South African government was heavily involved in the colonial-style military insurgency in South West Africa (Turner 1998). The international complications of the insurgency were such as to lead the government into military adventures which ultimately sapped the willingness of the White electorate to continue the war. Politically the formation of the Ovambo People's Congress in 1957 by Andimba Toivo Ya Toivo and a group of Cape Town-based workers may be taken as marking the beginning of the internal challenge to continued South African rule.

The move towards the use of political violence and a more unified African opposition began with the Old Location massacre in Windhoek in 1959, when the Ovambo People's Organization and other parties demonstrated against forced removals to the new ethnically defined neighbourhoods outside the city. This event provided the impetus for the unification of the various ethnically based organizations to form the South West African People's Organization (SWAPO). In 1966 SWAPO began an armed insurrection against the South African administration, having organized an external base in Tanzania. Thus it joined the group of diverse southern African liberation movements which were seeking the overthrow of the governments of South Africa, South West Africa, Rhodesia, Angola and Mozambique.

The South West African situation was profoundly affected by the termination of Portuguese rule and the independence of Angola in 1975. The struggle for power which ensued in the Angolan civil war was reflected in the conflict in South West Africa. Lines between the two were blurred, as SWAPO and the Angolan government fought the South African government and the UNITA (União Nacional para a Independência Total de Angola) movement. Indicative of the change was the removal of the SWAPO political headquarters to Lusaka and its military headquarters to Lubango (formerly Sa de Bandiera) in southern Angola.

In 1975 the South African army invaded Angola and occupied large sections of the country, in co operation with UNITA. However, international pressure and logistical problems resulted in withdrawal at the end of the year and a severe loss of South African military prestige. A more limited buffer zone was maintained to prevent SWAPO forces reaching northern South West Africa (Figure 7.8) (Barnard 1982).

The militarization of South West Africa began with the raising of new forces. In contravention of the terms of the original mandate, military bases were established in the territory and local inhabitants were recruited into the South African army. The first group, the Ovambo Battalion, was followed by the formidable counter-insurgency force, Koevoet. In addition, Angolans were trained as part of the South African army in a series of bases situated in the Caprivi Strip. The main aim was to keep the war firmly on the Angolan side of the border rather than allow it to develop in South West Africa itself. However, sporadic fighting took place in Ovamboland and the northern sectors of the White farming area.

DESTABILIZATION

The war in South West Africa became part of a wider struggle involving the other states in southern Africa. Moral and practical support for the African National Congress, Pan Africanist Congress, and SWAPO in the neighbouring states was such that the South African government sought to exert pressure on those states to prevent their territory being used as springboards for political and military actions. The states were particular objects of Pretoria's foreign policy after the assassination of Dr Verwoerd, through the 'outward policy' (1967–74) and 'detente' (1974–8) under Prime Minister John Vorster, and finally during destabilization (1979–89) under President P.W. Botha (Lemon 1991b). The 'total strategy' employed by the South African government in the 1980s involved financial and trade inducements as well as military destabilization in the hope of establishing a 'constellation of southern African states' under South African hegemony. The reach of the South African government extended from Lesotho to the bizarre attempt to overthrow the

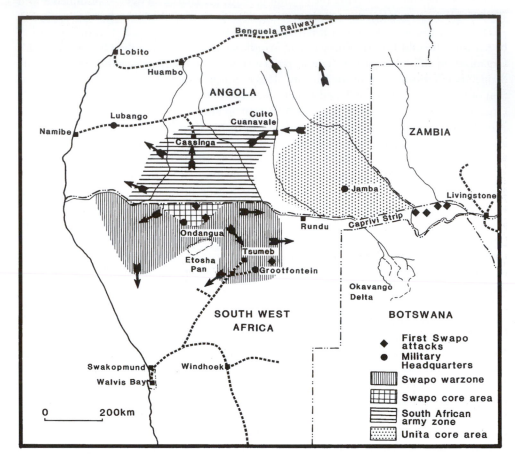

Figure 7.8 Extent of insurgency
Source: Modified after W.S. Barnard (1982) 'Die geografie van 'n revolusionere oorlog: SWAPO in
Sudwes-Afrika', *South African Geographer* 10: 157–74

government of the Seychelles in 1981, and the political assassination of the African National
Congress envoy, Dulcie September, in 1988 in Paris.

Destabilization began with low-level operations including the naval raid on Dar es Salaam
in 1972 (Stiff 1999). It became more overt and intense with the South African intervention
in the Angolan civil war at various stages between 1975 and 1989, and the effective arming
of the Renamo (Resistência Nacional Moçambicana) insurgency in Mozambique after 1980
(Figure 7.9). In both cases pressure was exerted on the new Marxist governments to prevent
them aiding SWAPO and the African National Congress. In Mozambique the South African
Defence Force inherited the Renamo operation from the Rhodesians who had assisted it in
the late 1970s. Through classic insurgency tactics Renamo effectively prevented the
Mozambican government from gaining full control over its territory, and contributed to the
national bankruptcy brought about by the adoption of doctrinaire communist principles.
Large-scale refugee movements led to further disruptions. In Angola the ability of the South
African army to prevent the defeat of UNITA, and the widespread destruction of the civil

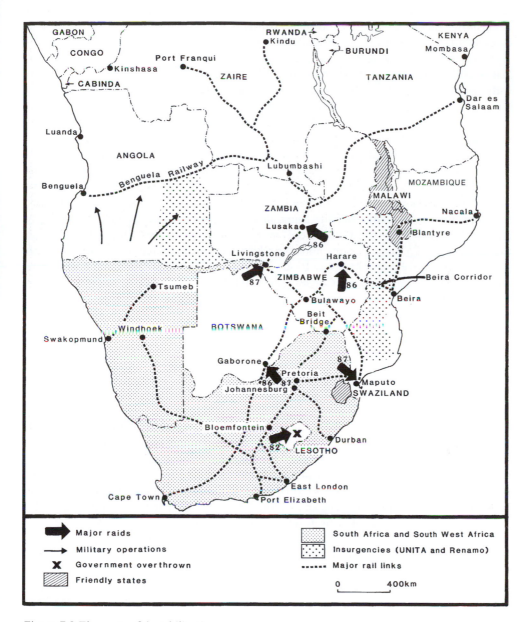

Figure 7.9 Elements of destabilization
Source: Compiled from various newspaper sources

war, had similar results. The wars in Angola and Mozambique were remarkable for the widespread use of land mines as a weapon against both armed forces and civilian areas. The legacy of the massive mining campaigns is likely to continue for decades to come, as demining is a slow and expensive process.

In 1984 the South African government entered into agreements with the Angolan government in Lusaka, and the Mozambican government through the Nkomati Accord. The agreements provided for the end of destabilization and the withdrawal of South African troops from Angola in return for assurances that the two countries would not be used as bases for attacks on South Africa or South West Africa. The diplomatic success was shortlived as South African destabilization activities continued, as did SWAPO and the African National Congress actions in Angola. Thus the nominal completion of the South African withdrawal from Angola in 1985 coincided with the capture of a South African commando unit involved in a sabotage mission against the Cabinda oil wells in northern Angola.

Destabilization was also directed towards other southern African states and towards the sabotaging of negotiations which were deemed unfavourable to the South African government. Lesotho was particularly susceptible to pressures. A major raid on Maseru in 1982 was aimed specifically at destroying the African National Congress presence in the country. In 1986 the Lesotho government was overthrown as a result of pressures exerted by South Africa through a slowdown in border control formalities, resulting in the paralysis of trade and population movements over the common border. A more amenable regime was duly installed. Raids on targets in Gaborone, Harare and Lusaka in 1986 effectively ended the Commonwealth Eminent Persons initiative which had been seeking a solution to South Africa's internal problems (Ramphal 1986). In 1987 raids on Gaborone and Livingstone indicated that African National Congress personnel would be targeted anywhere in the subcontinent. Militarily at least, South African destabilization succeeded in preventing overt action by its neighbours and keeping hostile bases some distance from South Africa's borders. However, the heavy financial and political costs of the policy were such that the government reverted to peace initiatives in 1988.

THE SOUTH AFRICAN DEFENCE FORCE

Destabilization was made possible by the numerical and technical dominance of the South African Defence Force in the subcontinent. After 1948 the permanent military establishment had initially been viewed with suspicion by the National Party, as many Afrikaner nationalists had actively opposed the South African entry into the Second World War, and some had supported the cause of National Socialist Germany. In addition the party was opposed to the arming of Africans and Coloureds, believing that this would weaken the position of the Whites, who alone were allowed to bear arms.

However, the perceived threat from countries elsewhere in Africa and the imposition of an arms embargo by Great Britain and the United States in 1963 led to a reassessment of the country's armed forces (Seegers 1996). The deterioration of the security situation in South West Africa necessitated an enlargement of the army, to augment the police force. The military call-up period for White men was extended to two years by 1977 with substantial tours of duty in the reserves thereafter. The length of these duties was such that 7 per cent of all White males aged between 18 and 45 were engaged in military service by 1980. This was not a high proportion if compared with Israel or a number of militarized states elsewhere in the world, but it was high for Western democracies, with which the White population of South Africa usually identified. This was to be one of the major causes of dissatisfaction with the National Party amongst the White electorate by the late 1980s.

In a major departure in principle, the government decided to arm the Coloured volunteers to the Cape Corps in 1973. In addition African units were raised for the Transkei in 1974, to be followed by other homeland units thereafter. African units in the South African Defence Force were initiated in 1974 in South West Africa and in South Africa soon afterwards. By 1980 approximately half the army was no longer White. Ten years later the proportion had risen to over three-quarters.

In terms of size the South African Defence Force was larger, and better trained, than any other in the region (Figure 7.10). It was capable of being augmented with reserves in times of crisis to provide overwhelming numerical superiority. Consequently by the late 1970s large numbers of men could be deployed for long periods of time in South West Africa or in the African townships of South Africa. It is worth noting that by 1980 SWAPO probably had no more than 2,000 men in the field at any given time, yet it required 20,000 to 25,000 members of the South African Defence Force to counter them.

CONSTELLATION OF SOUTHERN AFRICAN STATES

The South African government pursued destabilization and open warfare as part of a policy to create security. An equally significant aspect was the concept of the 'constellation of southern African states' bound together economically, which pursued politically different, but not antagonistic, courses (Malan 1980). In part this was related to the thinking that the way to international acceptance was through African co-operation.

The concept of a southern African commonwealth was first suggested by Dr H.F. Verwoerd in 1963 as an adjunct to the state apartheid programme. It was his successor Mr B.J. Vorster who actively pursued an outward policy, aimed at building a ring of friendly states between South Africa and the remainder of the continent. South Africa, Rhodesia and the Portuguese colonies drew closer together both economically and diplomatically as outside hostility increased. Practical co-operation included such major schemes as the Cahora Bassa hydro-electric scheme and the Ruacana Water scheme (Figure 7.11). In both cases the Portuguese colonies were partially integrated into the South African economy and security system. However, one of the major casualties of the subsequent destabilization era was the destruction of the power lines from Cahora Bassa to the eastern Transvaal.

Only one independent state in southern Africa co-operated fully with the South African government in its outward policy, namely Malawi. President Kamuzu Banda sought aid from wherever he could. South African finance thus became available for a variety of projects, of which the building of the new capital city at Lilongwe was the most spectacular.

The independence of Angola and Mozambique in 1975 caused major shifts in South Africa's foreign policies, as in much else, as the cordon sanitaire around the country was effectively breached and hostile states now shared common borders with it. The crisis in Rhodesia became more pressing as the Mozambican government closed the border between the two states and Rhodesian trade was redirected through Botswana. An emergency railway link was subsequently built through Beit Bridge to connect the two countries directly.

The concept of a constellation of southern African states was revived by Mr P.W. Botha in 1977. In the following year the Development Bank of Southern Africa was established to offer South African aid in funding suitable projects in the subcontinent. Nevertheless, the Rhodesian war prevented any meaningful detente in southern Africa. The ending of South African support for the UDI government in Rhodesia cleared the way for the establishment of the grouping. However, the subsequent success of ZANU-PF in winning the first elections in an independent Zimbabwe effectively ended the concept of a South African-led southern African grouping.

Following the independence of Zimbabwe in 1980 the various frontline states in southern Africa formed the Southern African Development Coordination Conference (SADCC) at a

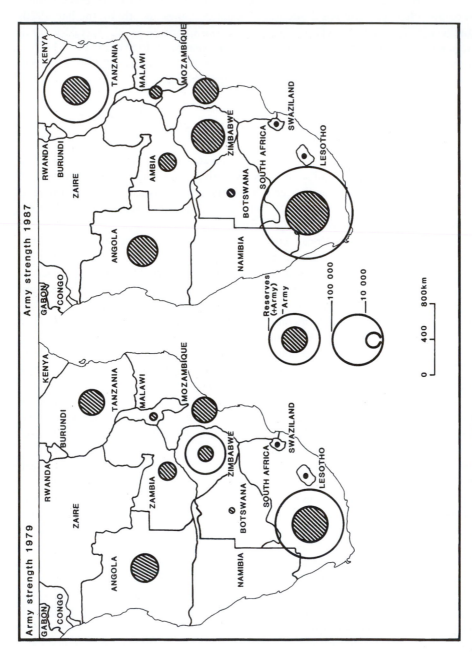

Figure 7.10 Regional army strengths, 1979 and 1987

Source: Based on statistics in *Statesman's Yearbook* for 1979 and 1987

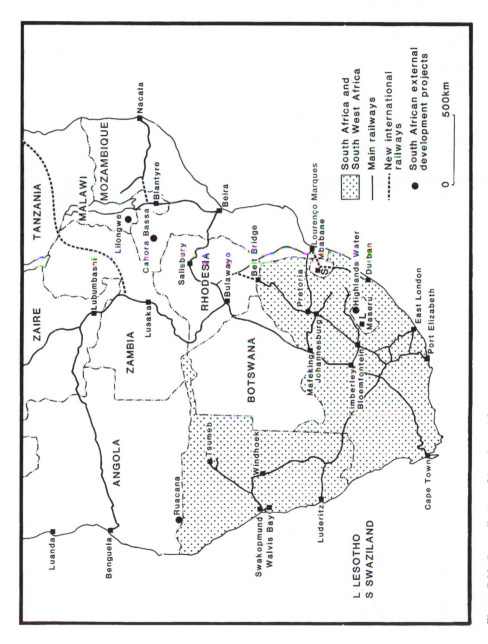

Figure 7.11 Constellation of Southern African States
Source: Compiled from various newspaper sources

meeting in Lusaka (Gibb 1988). The express purpose of the organization was the lessening of the states' economic dependence upon South Africa. The nine countries, increased to ten with the independence of Namibia ten years later, in particular sought to decrease their dependence on the South African transportation system. In 1980 South Africa accounted for 76 per cent of the Gross National Product of the region, and was the major trading partner of all but Tanzania and Angola (Figure 7.12).

The particular problems of Botswana, Lesotho and Swaziland as members of the Southern African Customs Union were evident (Gibb 1987). The other countries exhibited greater independence, although in 1980 the Zimbabwean economy was heavily dependent upon the South African economy as a result of the international trade sanctions imposed upon Rhodesia after the unilateral declaration of independence in 1965.

MINE LABOUR

The South African mining industry developed a wide recruiting zone in southern Africa based on the principle of employing single African men as temporary migrants. Although Europeans were recruited as permanent immigrants, successive laws sought to prevent Africans obtaining rights of abode in the country, culminating in the Aliens Control Act of 1991 (Peberdy and Crush 1998). After the Anglo-Boer War (1899–1902) it was the recruitment of labourers from Mozambique which effectively supplied the manpower to re-establish the mines and boost production. Health problems with those recruited from

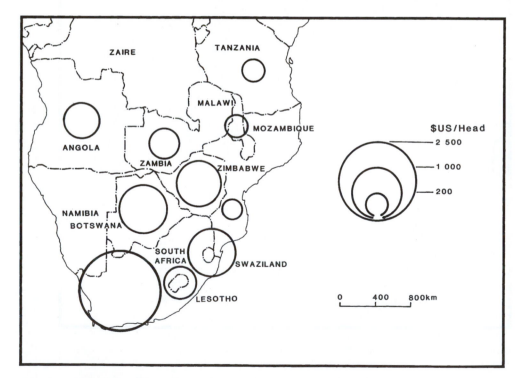

Figure 7.12 Southern Africa: Gross National Product per head, 1984
Source: Based on statistics in *Statesman's Yearbook 1984*

central Africa led to restrictions on employing men from north of 22°S until 1932, while the governments of Southern and Northern Rhodesia attempted to retain as much labour for their own economies as possible, and thus discouraged workers from going to South Africa. The scale of this demand was such that the local supply of labour remained inadequate until the 1970s and an intensive search to gain men from other parts of the subcontinent ensued. After the 1970s the employers' need to diversify the sources of supply, to reduce reliance on any one source, resulted in the continued employment of foreign labour.

The National Party policies initially had little impact upon mine labour recruitment (Figure 7.13). Thus the patterns of supply in 1960 essentially reflected those of 1948. Changes came with African independence, as the more northerly countries halted recruitment to demonstrate their opposition to the migratory labour system and the South African government. However, the mining houses' demands for more labour in the prosperous 1960s resulted in a boost in numbers and the development of Malawi as the main source of foreign manpower for the South African mines. In 1974 an air crash in Botswana resulted in the death of seventy-five Malawian migrant miners, and recruitment was suspended. The civil war in Mozambique resulted in a severe curtailment of the supply from that country. As a result the number of foreign miners in South Africa fell from an average of 316,000 in 1973 to 196,000 in 1977, at which level it remained fairly constant thereafter.

In 1978 the Southern African Labour Commission was established by the countries supplying labour to South Africa with the aim of eliminating the migratory system (Crush 1991). However, little action was taken. Only Zimbabwe completely disengaged after independence, thereby fulfilling the aims of earlier colonial governments. The system of deferred payments, whereby part, sometimes as much as 60 per cent, of the migrants' earnings were remitted to the government concerned, gave the supplying governments a vested interest in its continuance. Thus, far from being able to exert pressure on South Africa as envisaged, the South African recruitment agencies were able to exert pressure on the source countries.

The reduction in the number of foreign workers was made up with local labour and men recruited from Lesotho, which by 1976 had become the largest supplier of labour to South Africa's mines. Mine labourers thus migrated over shorter distances than before: this enabled many to return home at weekends, placing them in the same position as homeland migrants.

STRATEGIC INDUSTRIES

Economic sanctions proved to be the most effective weapon of the international community against apartheid. A voluntary arms embargo was imposed by the United Nations in 1963. Great Britain and the United States ceased their trade, but their place was taken by France and other countries. In 1977 the arms embargo became mandatory, but by this time the country had become largely self-sufficient in the production of those weapons required for the types of war waged by the South African Defence Force.

Strategic projects, including the development of SASOL (South African Coal, Oil and Gas Corporation), were initiated after 1948. SASOL was created at what became Sasolburg and was enhanced with the construction of Secunda in the 1970s and 1980s for the extraction of oil from coal. The aim was to establish strategic self-sufficiency in oil production. The Mossel Bay oil and gas project in the 1980s similarly sought to lessen South Africa's dependence upon imported oil. However, its cost exceeded R10 billion, with little chance of ever showing a profit. The government also invested in a nuclear power station at Koeberg in the

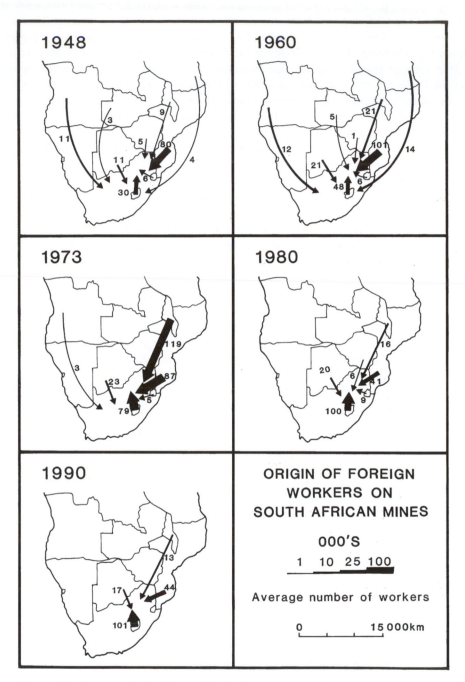

Figure 7.13 Sources of mine labour, 1948–90
Source: Compiled from statistics in the *Annual Reports* of the Chamber of Mines

western Cape, together with the atomic research station at Pelindaba near Pretoria. Out of this was to come one of the most fearsome products of the apartheid era, the atomic bomb. It was only in 1993 that the existence of six atomic bombs was revealed and they were duly dismantled before the advent of the new government.

Other such projects included the investment in local content for motor vehicles, more especially the diesel engine plant at Atlantis, together with the armament industry, including the manufacture of aircraft and coastal vessels (Figure 7.14) (Rogerson 1990). Armscor (originally the Armaments Production Board) was formed in 1964 to counter the United Nations arms embargo with a programme of strategic industrial development. The corporation was involved, through a number of subsidiaries, in the production of a wide range of military equipment ranging from aircraft production at Kempton Park to explosives and projectiles in the western Cape. In addition Armscor contracted out a wide range of products to private firms. The main areas of activity were located in the Witwatersrand and Pretoria. Government subsidies for these schemes were such that no economic return could be expected and ultimately they helped to undermine the South African economy.

Mineral export promotion projects were also developed. Coal exports from the eastern Transvaal and iron ore from the northern Cape were designed to add to South Africa's

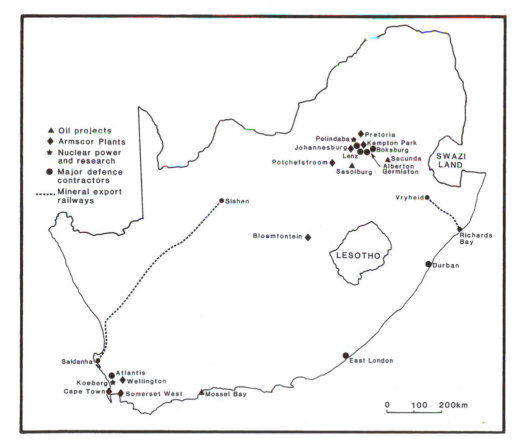

Figure 7.14 Strategic industries
Source: Various sources, including C.M. Rogerson (1990) 'Defending apartheid: Armscor and the geography of military production in South Africa', *GeoJournal* 22: 241–50

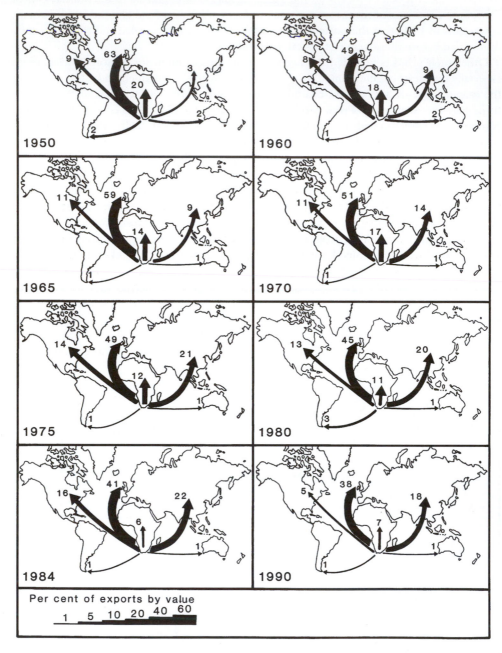

Figure 7.15 Exports from South Africa, 1950–90, by continent
Source: Compiled from statistics in the *Statistical Yearbook of South Africa*, various dates, and the *Monthly Abstract of Trade Statistics*, Pretoria: Government Printer

foreign earning capacity. Special railways from the mines to the newly constructed ports of Saldanha and Richards Bay were among the strategic schemes undertaken in the 1970s.

TRADE SANCTIONS

In other spheres trade sanctions were irregular in imposition and effect (Ramphal 1989). India first imposed sanctions in 1946 in response to the anti-Indian legislation of the same year. Further countries banned specific forms of economic activity, such as direct investment by Japan in 1964, which did not preclude the expansion of trade between the two states. Sweden, a leading political opponent of apartheid, only forbade new investment in 1979.

It was in response to the riots of 1984 that comprehensive measures were widely adopted against South Africa. The United Nations Security Council, the European Community, the Commonwealth and the United States all imposed voluntary embargoes within the next few years. Probably the most influential was the enactment of the American Comprehensive Anti-Apartheid Act of 1986. Credit squeezes, oil embargoes and the restriction of precious and strategic metal exports all followed. Disinvestment became serious and trade, particularly American trade figures, fell markedly. The South African economy was thus placed under substantial pressure, reducing economic growth rates in the late 1980s to near zero, while the population continued to expand at approximately 2.0–2.5 per cent per annum. The countries reducing their imports from South Africa were led by the United States, whose trade halved between the early 1980s and 1990 (Figure 7.15). Other countries, notably Germany, Japan and Taiwan, made up a great deal of the shortfall. External pressures thus contributed significantly to the final demise of apartheid.

8   Dismantling apartheid

Dismantling apartheid is an on-going process that will take far longer to achieve than the nine-year transitional period surveyed in this chapter. In the period between 1990 and 1999 substantial progress was made in removing many of the more openly discriminatory measures of the apartheid era. However, other issues remained essentially conditioned by the apartheid mould as a legacy for the new millennium. Some of these are outlined in Chapter 9. Transition to the post-apartheid era began with the speech delivered to parliament by President F.W. de Klerk on 2 February 1990, in which he announced the unbanning of the African National Congress, the Pan Africanist Congress, the South African Communist Party and other organizations. This was accompanied by the release of political prisoners, notably the African National Congress leader, Nelson Mandela.

TRANSITION

A convoluted series of negotiations were then undertaken by the government and the various internal and external political parties and organizations to address the problems facing the country. At the same time the levels of violence and destabilization reached new heights as the various contenders sought to gain political advantage. Significant changes of attitude had taken place within the governing National Party, as the newly elected President F.W. de Klerk attempted to modernize apartheid more radically than his predecessor. Equally, the African National Congress leadership had reconsidered its options and issued the Harare Declaration in 1989, which opened the way for a possible negotiated settlement rather than the armed seizure of power as advocated at the Morogoro (1969) and Kabwe (1985) consultative conferences (McKinley 1997).

Eventually the Convention for a Democratic South Africa (CODESA) was convened in December 1991 to draw up a new constitution for the country (Davenport 1998). The nineteen participating delegations included the homeland governments as well as the internal political parties and external liberation movements. Bitter acrimony surrounded the two sessions of the Convention, which broke up in disarray the following year and the country was subjected to a campaign of 'rolling mass action' in favour of revolutionary constitutional changes.

At the same time the ruling Afrikaner elite accepted the inevitable loss of power. Even the Afrikaner Broederbond changed its name and admitted women and sympathetic members of all population groups in 1992. Recognizing the fundamental shift in its negotiating position, in 1992 the government called a referendum of the White electorate to gain support for its position of creating a united, democratic, South Africa. This it did by a two-thirds majority in what was to be the last all-White election. The electorate of only one of the fifteen polling districts did not support the government and indicated a desire to pursue the concept of apartheid (Figure 8.1): the dissident Northern Transvaal possessed a 97 per cent African population majority and a small White electorate which was fearful of change.

In 1993 the Multi-Party Negotiating Forum, with twenty-six participating delegations, met at the World Trade Centre at Kempton Park outside Johannesburg. It drew up the interim constitution and agreed upon the establishment of a five-year Government of National Unity, during which time a measure of power sharing between the majority and minority parties would be maintained. It was a measure of the extent of the compromises reached that the country was to have eleven official languages, promoted by the Pan-South African Language Board, although English effectively became the lingua franca for official transactions.

The negotiations were conducted against a background of violence and symbolically it was an act of violence, the assassination of the South African Communist Party leader Chris Hani, in 1993 which focused the minds of the negotiators to find a solution to the impasse, fix an election date and bring the country back from the prospect of civil war. In the five-year period (1990–4) some 16,000 people died in political violence, almost half of them in KwaZulu and Natal. The struggle for power between the African National Congress and the Inkatha Freedom Party was inflamed by covert government intervention to prevent the

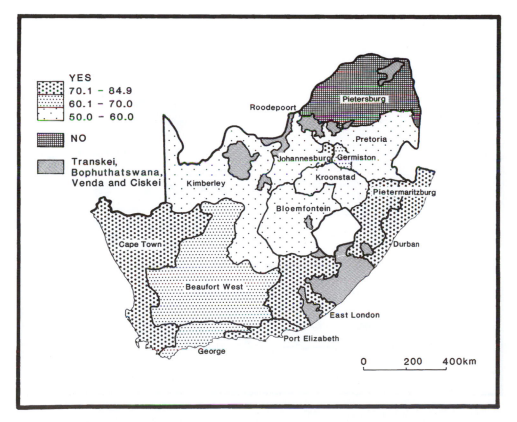

Figure 8.1 Referendum results, 1992
Source: Based on results published in *Eastern Province Herald*, 19 March 1992, Port Elizabeth

latter from being defeated. The scale of the conflict varied from individual killings to the employment of systematic massacres, usually of the elderly, women and children (Figure 8.2). The development of this new dimension in the conflict resulted in the perpetration of some ninety-four massacres (of ten or more people at a time) between January 1990 and April 1994, compared with thirty-three between 1948 and 1989. Local disputes, over land and livestock, notably in the KwaZulu homeland, were caught up in the political conflict, and continued after the elections, resulting in the creation of an all-pervasive atmosphere of fear (Gutteridge and Spence 1997).

The violence in Natal and KwaZulu was paralleled in the Witwatersrand region, where conflict between Zulu hostel workers and township residents, and other cleavages in society, added to the death toll. The Boipatong massacre by Inkatha hostel workers near Vanderbijl-park, in the Vaal Triangle, in 1992 resulted in the death of forty-six people and set the tone of the ensuing conflict (Coleman 1998). Ultimately the killing of fifty people in the course of an Inkatha march past Shell House, the African National Congress headquarters in Johannesburg, in March 1994 resulted in the imposition of a state of emergency in KwaZulu. It is indicative of the measure of political control over the killings that although 305 people were

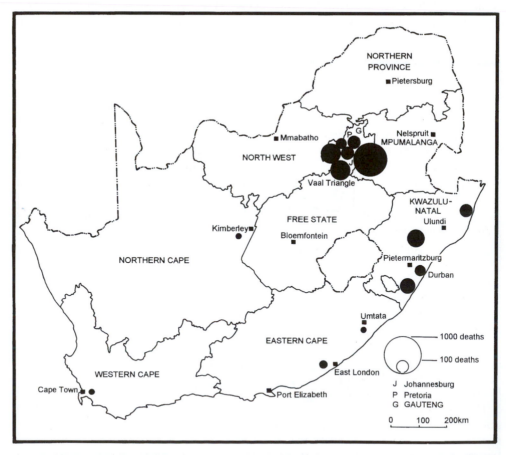

Figure 8.2 Lives lost in massacres, 1990–6
Source: Based on information in M. Coleman (1998) *A Crime Against Humanity: Analysing the Repression of the Apartheid State*, Cape Town: David Philip

killed in April 1994, mostly before the African National Congress–Inkatha Freedom Party accord, the number was reduced to 125 in the following month. Other campaigns of murder and massacre were carried out by the Azanian People's Liberation Army, aimed at a violent revolutionary overthrow of the government, and various government death squads, aimed at destabilizing opposition groups.

The independent homeland structures were also undermined. In 1988 the civil government of Transkei had been overthrown by the Transkei army. However, an attempted *coup d'état* in Bophuthatswana was suppressed by the intervention of the South African army. The new Transkei regime supported the African National Congress and Pan Africanist Congress, providing military training bases, and it was therefore the object of one of the last cross-border raids by the South African Defence Force in 1993. The Venda and Ciskei governments were overthrown by their armies in 1990, in favour of reincorporation into South Africa. Subsequently the new Ciskei military leader became hostile to change, which he demonstrated forcibly in a confrontation with an African National Congress-led march on Bisho in 1992 in which twenty-nine people were killed. However, the administration of the state was taken over by South Africa just before the elections as the homeland government disintegrated. The same fate overtook the Bophuthatswana government immediately before the 1994 elections after a failed White right-wing military intervention in support of its attempted retention of independence.

UNDOING ADMINISTRATIVE APARTHEID

The new constitution provided for the redrawing of the administrative map of South Africa (Figure 8.3). All the homelands were abolished and reintegrated into the administrative structures of the country (Christopher 1995b). Nine new provinces were created, based essentially upon the nine planning regions established in 1982 (see Figure 2.6, p. 56). During the constitutional negotiations various configurations were proposed and a compromise reached over the division of the Cape Province. The resultant Northern Cape was the largest province in area, but by far the smallest in terms of population. The core province of the Pretoria-Witwatersrand-Vaal Triangle (PWV) housed a substantial population (seven million) and a disproportionate (40 per cent) share of the National Gross Domestic Product. Elsewhere the negotiators had sought to link the impoverished homelands to neighbouring White areas, in the expectation that there would be a degree of equalization in the provision of services within a province, and therefore economic and social improvement of the former homelands.

The provinces were designed to be a second tier of government, with elected councils and administered by premiers. However, the concept was not essentially federal as the central government retained control of their finances and in practice the premiers were appointed by the national party political leaderships. The concept of a semi-independent Kingdom of KwaZulu within the Republic of South Africa was shelved, leading to continuing acrimony between the parties concerned.

The redrawing of the provincial boundaries was sanctioned in the constitution, but the government was loath to upset the delicate balances achieved in the constitution and confirmed in the subsequent national and provincial elections. Thus the anomalous position of the detached Umzimkulu district in the Eastern Cape was not resolved either by reincorporating the intervening Xhosa-speaking East Griqualand area into the province, or including Umzimkulu into KwaZulu-Natal. Similarly, the more violent Bushbuck Ridge dispute

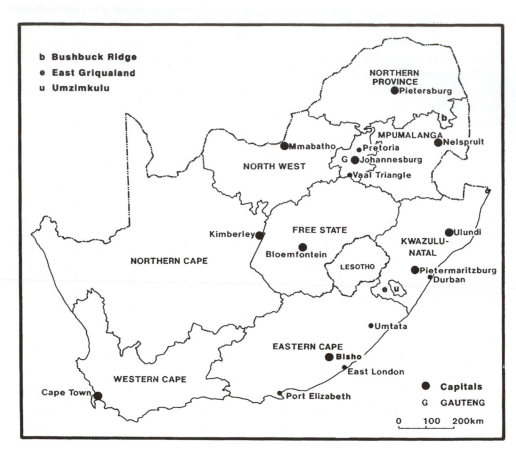

Figure 8.3 New provinces and capitals, 1994
Source: After A.J. Christopher (1995) 'Regionalism and ethnicity in South Africa 1990–1994', *Area* 28: 1–11

between Mpumalanga and Northern Province was not resolved by the compromise that the area should remain in the latter province, but services should be provided by the former, on an agency basis (Ramutsindela 1998). Other disputes were contained on a less confrontational basis.

So complete an eradication of the homelands was achieved, that only the name 'KwaZulu' survived, though hyphenated to the old provincial name, 'Natal'. Equally the historic settler name 'Transvaal' disappeared from the map as the Eastern Transvaal province was renamed Mpumalanga ('where the sun rises'), and the Northern Transvaal was shortened to Northern Province. The cumbersome 'PWV' was appropriately renamed Gauteng ('place of gold'). Even the Orange Free State lost its titular historic link with the Royal House of the Netherlands, becoming just the Free State Province.

The choice of the provincial capitals proved to be less contentious than that of the homeland capitals in an earlier era. The metropolitan city of Johannesburg became the capital of Gauteng. In some cases continuity was maintained with the selection of the White provincial capitals of Cape Town and Bloemfontein, and the homeland capitals of Mafikeng and Bisho. No viable alternatives to Pietersburg in the Northern Province or Kimberley in the Northern Cape were proposed and Nelspruit was selected for Mpumalanga. Problems arose with the amalgamation of KwaZulu and Natal, resulting in a compromise solution whereby Pietermaritzburg and Ulundi became joint capitals. The other homeland capitals, including Umtata, were reduced to local administrative centres. Pretoria retained the status of the national capital, and there was significant state support for moving the seat of parliament from Cape Town in order to end the colonial compromise of a split national capital.

The Freedom Front, which had been formed to diffuse the Afrikaner White right-wing threat to disrupt the elections in 1994, still sought an Afrikaner homeland. A Volkstaat Council was established with the task of inquiring into the implications of the principle of self-determination enshrined in the constitution. However, the possibilities of establishing a territorial homeland receded, although several ingenious maps of a potential Afrikaner state were produced between 1994 and 1997 (South Africa 1997). Only one Afrikaner majority area, based on Pretoria, could be identified, through the judicious division of towns and cities to exclude the African majorities (Figure 8.4). Three further areas, including Orania, were identified, but the voluntary population shifts necessary for their fulfilment were not feasible. Ultimately the Volkstaat Council recognized that promoting Afrikaner cultural and linguistic values through community councils appeared a practical solution to the basic

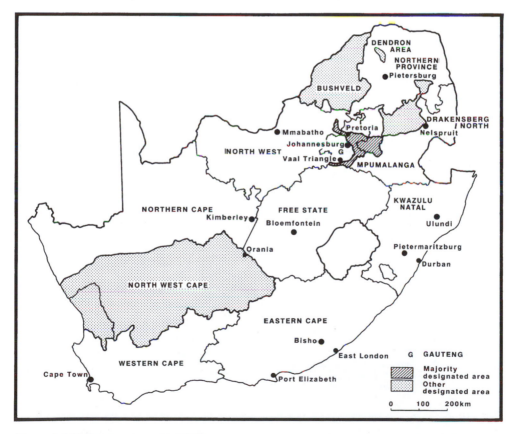

Figure 8.4 Volkstaat proposals, 1997
Source: After South Africa (1997) *Self-determination for Afrikaners,* Pretoria: Government Printer

dilemma which had initially prompted the policy of apartheid, namely the preservation of the Afrikaner nation.

The pattern of local government also underwent significant reform during the course of the multi-party negotiations, with the enactment of the Local Government Transition Act of 1993. The existing racially defined councils were suspended and replaced by negotiating forums composed of members of the existing structures and hitherto unrepresented bodies such as the civic associations on a fifty-fifty basis (de Beer and Lourens 1995). It is worth noting that by this stage any semblance of elected African local authorities had ceased to function for lack of legitimacy. The Transitional Local Councils came into being as agreements in the forums were reached.

Some forty-two new local councils were constituted, which grouped together urban and rural areas in an attempt to create viable administrations capable of delivering the services promised to the citizenry (Figure 8.5) (Delport 1997). In addition six metro-politan areas were designated, with structures befitting the complexity of their govern-ment (Maharaj 1997). Each was subdivided to prevent the administration becoming unwieldy and unresponsive to the electorate. Thus the Cape Metropolitan Council co-ordinated the policies of six separate municipalities. It is notable that all traces of the

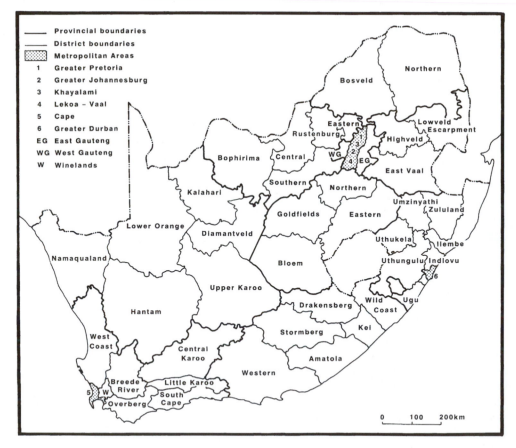

Figure 8.5 Local government structures, 1997
Source: Based on information in A. Delport (ed.) (1997) *Official South African Local Government Yearbook 1997/8*, Johannesburg: Gaffney

former homelands were erased as both district and metropolitan councils overlapped the former boundaries.

The more than 800 Transitional Local Councils experienced substantial financial and other problems as a result of a country-wide boycott of the payment of rates and levies and the burden of extending the provision of services to communities previously neglected. As a result, the central government indicated that it would significantly restructure the councils, reducing their numbers and increasing their powers. Elections for the new councils were scheduled for 2000, thereby completing the political transformation of the country.

ELECTING A NEW GOVERNMENT

Elections for the National Assembly and the Provincial Councils were held under universal franchise on 27 April 1994 (Johnson and Schlemmer 1996). No voters' roll could be compiled in the time available and the system of proportional representation was adopted as a means of overcoming the gerrymandering associated with the constituency system of the previous parliament. In the transition period before the elections (1993–4), the country was run by a Transitional Executive Council to oversee the process. The elections themselves were supervised by a neutral Independent Electoral Commission.

A wide range of parties contested the election. The two dominant parties were the African National Congress (ANC), leading a tri-partite alliance with its partners the Congress of South African Trade Unions (COSATU) and the South African Communist Party, on the one hand, and the National Party (NP) on the other. The Inkatha Freedom Party (IFP) only joined the process one week before the election, as a result of further constitutional and land concessions extracted in return for the promise of an end to political violence. Boycotts were conducted by the White Conservative Party and the Afrikaner Resistance Movement (AWB) at one extreme, and by the Azanian People's Organization at the other. It is maybe a measure of the broad political accommodation of the era that the youth wing of the Pan Africanist Congress popularized the anti-White slogan 'one settler, one bullet', yet participated in the democratic process. A host of other parties, promoting Christian or Moslem religious values, workers' rights, green issues, federalism and liberal democracy, were joined by a few highly idiosyncratic organizations such as the KISS (Keep It Straight and Simple) Party and the SOCCER (Sports Organization for Collective Contributions and Equal Rights) Party (Reynolds 1994).

Despite the forebodings concerning White right-wing disruption, the elections were conducted in peace and calm with a festive atmosphere demonstrating evidence of considerable national goodwill. Some 19.5 million of the estimated 22.6 million people entitled to vote cast their ballots, a remarkable achievement in cooperation. Although some doubts were cast upon the accuracy of the results, their general acceptability outweighed other considerations. The African National Congress, as the foremost liberation movement, won a resounding victory, gaining 62.6 per cent of the votes cast, and so acquired 252 of the 400 seats in the National Assembly. This was just short of the two-thirds majority required to write the final constitution unilaterally. The National Party gained 20.4 per cent, just above the one-fifth required to nominate an Executive Deputy President in the Government of National Unity. The Inkatha Freedom Party received 10.6 per cent of the vote, thereby securing three seats in the national cabinet. The minor parties fared badly, with only four other parties securing any seats in the National Assembly.

The election results exhibited some marked regional variations (Figure 8.6) (Christopher 1996). The African National Congress secured outright majorities in six of the nine provinces, and narrowly failed to do so in a seventh (Northern Cape). The scale of the victory was overwhelming in the non-Zulu-speaking African majority areas. Thus in the Northern Transvaal the African National Congress secured 91.6 per cent of the vote. It should be reiterated that this was the region which in the 1992 all-White referendum had rejected the reform process. No better illustration of the manipulation of the franchise qualification could be presented. High levels of support in the Eastern Cape, North West, Eastern Transvaal and Orange Free State reflected the demographics, where the Pan Africanist Congress failed to challenge the African National Congress.

The National Party secured the majority of the vote in the Western Cape and over large parts of the Northern Cape. This reflected the party's success in gaining the support of the majority of Afrikaans-speaking Coloured voters of the region, who were fearful of the impact of African rule. Thus the provincial council of the Western Cape was dominated by the National Party, which elected the sole White premier in the country. In KwaZulu-Natal, the Inkatha Freedom Party secured a bare majority of 50.3 per cent, but sufficient to enable it to

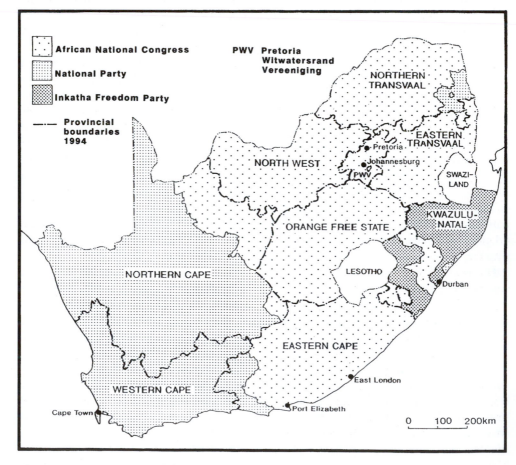

Figure 8.6 Election results, 1994
Source: After A.J. Christopher (1996) 'Regional patterns in South Africa's post-apartheid election 1994', *Environment and Planning C: Government and Policy* 14: 55–69

dominate the provincial government and obtain the premiership. It is notable that the African National Congress gained support in the urban areas of the province, but failed to make much impact in the more traditionally organized rural areas.

The National Party left the Government of National Unity following the enactment of the final constitution in 1996. In this manner it attempted to demonstrate that it was an opposition party intent on providing an alternative to the governing party. Auspiciously, the other party in the cabinet, the Inkatha Freedom Party, remained in the Government of National Unity and so the very high levels of violence which plagued the pre-1994 election period were not repeated in 1999.

The second universal franchise election in June 1999 was conducted according to a national voters' roll, although the system of allocating seats in the National Assembly remained that of proportional representation. The total number of votes cast declined to 15.98 million, reflecting the problems associated with the compilation of the voters' roll, notably the issue of identity documents and registration, and a degree of abstentionism provoked by disillusionment with the pace of change. The African National Congress won a resounding victory at national level with 66.4 per cent of the vote and 266 of the 400 seats in the National Assembly, only one short of the two-thirds majority. The opposition parties were greatly reduced in strength, with none gaining even 10 per cent of the vote, resulting in the striking spatial dominance of the governing party (Figure 8.7). The Democratic Party which gained 9.6 per cent of the vote became the official opposition, displacing the New National Party whose support slumped to only 6.9 per cent. Significantly, the Inkatha Freedom Party remained a coalition partner in the new government formed by President Thabo Mbeki.

At provincial level, in KwaZulu-Natal the Inkatha Freedom Party remained the largest party but had to form a coalition with the African National Congress in order to secure control of the provincial administration. Although the African National Congress gained the largest number of votes in the Western Cape, the New National Party retained control by forming a coalition with the smaller parties. A local anomaly in the Umtata region was the success of the United Democratic Movement led by the former Transkei military leader, General Bantu Holomisa. The United Christian Democratic Party of ex-President Lucas Mangope of Bophuthatswana similarly attained some success in the North West. On the other hand the Freedom Front gained under 1 per cent of the vote, indicating that the idea of a territorial Afrikaner homeland had finally been abandoned.

Local government elections for new transitional local councils and rural councils were conducted in 1995. Those in KwaZulu-Natal were postponed until 1996 as violence made the registration of voters and the demarcation of wards impossible in places. The same occurred in parts of the Western Cape where voter registration had been problematical. The system devised for the selection of the new councils was complex. Sixty per cent of the seats were elected on the basis of geographically defined wards, with the remaining 40 per cent of the seats elected by proportional representation. The ward boundaries were drawn in such a manner that half represented the former White, Coloured and Indian areas, and the other half represented the former African areas, irrespective of the numbers of voters involved. In this way the two major contenders in the negotiations were assured of representation. However, the internal demarcation of the metropolitan areas was the subject of complex negotiations (Lemon 1996).

The local government elections produced results similar to those of the 1994 national and provincial elections. In Pretoria the National Party and the African National Congress each won half the wards, indicating the continued impact of apartheid urban planning on the

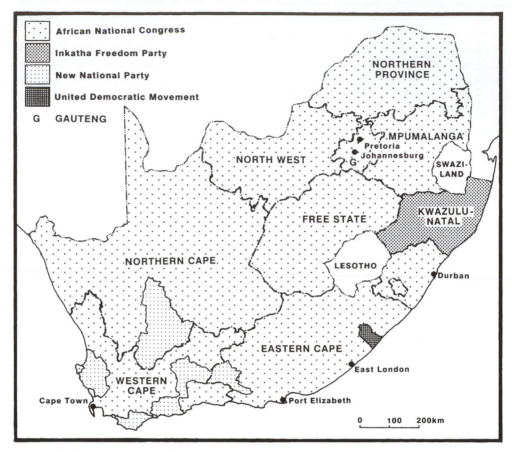

Figure 8.7 Election results, 1999
Source: Based on results of the Independent Electoral Commission, web site:
http://www.elections.org.za (9 June 1999)

outcome. In KwaZulu-Natal the African National Congress won control of the prosperous Durban Metropolitan area, while Inkatha was largely confined to the poor rural areas.

An examination of the Port Elizabeth results indicates some of the outcomes of the compromise involved in the local government system. A nominated Transitional Local Council had been installed in 1993, one of the first in the country, paralleling the collapse of the Ibhayi City Council. In 1995 an elected council of fifty-five members included twenty-one who were elected by proportional representation and thirty-four for geographically defined single-member wards, half in the former African areas and half in the former White, Coloured and Asian areas (Figure 8.8). However, as a result of the construction of new African housing schemes in the open buffer areas between former group areas, two wards in the nominal White-Coloured-Asian area were occupied predominantly by Africans. Both were won by the African National Congress. Thus at ward level the African National Congress secured 56 per cent (nineteen) of the seats, but 71 per cent (fifteen) of the seats awarded on the basis of proportional representation. The disparity indicated the extent of the gerrymandering involved in the delimitation.

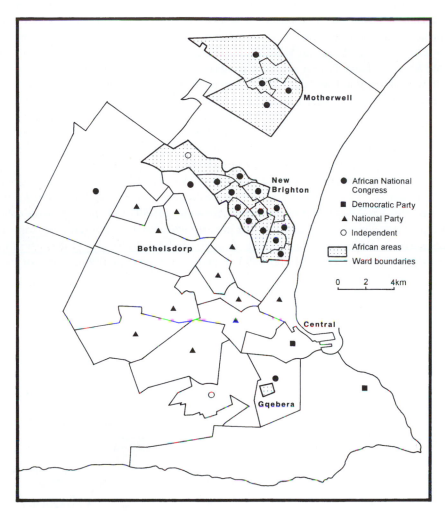

Figure 8.8 Municipal election results, 1995
Source: Based on information supplied by the Port Elizabeth Municipality

RECONSTRUCTION AND RESTITUTION

The African National Congress was elected as a liberation movement, whose tasks were eloquently summed up in the Reconstruction and Development Programme (African National Congress 1994), which aimed at redressing the social and economic imbalances created during the apartheid era while fostering nation building and democracy (Bond and Khosa 1999). The Programme became a major platform of the national government which directed key improvement policies and was supervised by a co-ordinating ministry in the initial phase (1994–6). The Presidential pilot projects were the most prestigious and clearly focused. Although many were of a country-wide nature, notably targeting improvements in the provision of education, health care, water supply, electricity, telephones and security, a number were place-specific aimed at alleviating problems in regions of particular deprivation (Figure 8.9).

The projects were highly varied and ranged from the major infrastructural schemes in Katorus (KAtlahong, TOkoza and VoslooRUS) in the violence-ravaged East Rand and the Integrated Serviced Land Project on the Cape Flats, to the smaller projects of upgrading services in several towns in KwaZulu-Natal. Some projects involved the complete replanning of the infrastructure, while others were more concerned with the provision of services, particularly water and sewerage. It might be emphasized that the projects were given a high profile and represented the launch of the programme in each province. As such they ran parallel with other national and local schemes with the same aims and eventually merged with the broader aspects of reconstruction (Khosa 1998).

The programme of land restitution and redistribution was begun in 1991. The Abolition of Racially Based Land Measures Act of 1991 repealed the Natives Land Act (1913), the Native Trust and Land Act (1936), the Group Areas Act (1950) and the numerous other foundations of spatially applied apartheid. The government also set about resolving some of the issues raised by this revolutionary move. It appointed the (Advisory) Commission on Land Tenure, which between 1991 and 1994 disposed of nearly 800,000 hectares of state

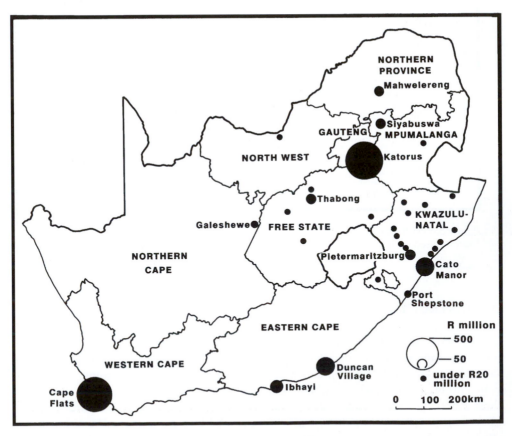

Figure 8.9 Presidential Reconstruction and Development Projects
Source: Based on information in *Hansard*, Interpellations 1998, cols. 626–632 Presidential Projects

land outstanding from the homelands consolidation programme (Figure 8.10) (Christopher 1995c). The Commission also began the process of resolving claims for the restitution of lands expropriated in the apartheid era, restoring some 222,000 hectares to sixty-three communities (South Africa 1995: 11).

The Government of National Unity established a more comprehensive, but cumbersome, structure to deal with land claims under the Restitution of Land Rights Act of 1994. A Commission on the Restitution of Land Rights, supported by a Land Claims Court, was charged with investigating and settling all claims by people who had lost land rights since the enactment of the Natives Land Act in 1913 as a result of racially based land legislation. The issue of acknowledging indigenous land rights dating back to colonial dispossession was not reopened as the new government wished to avoid the complex overlapping claims and apparently unending litigation which had attended the issue in other countries (Christopher 1997; McNeil 1989).

Nearly 64,000 claims had been lodged with the Commission by the closing date for submission at the end of 1998. Although most claims were for ownership of property, other

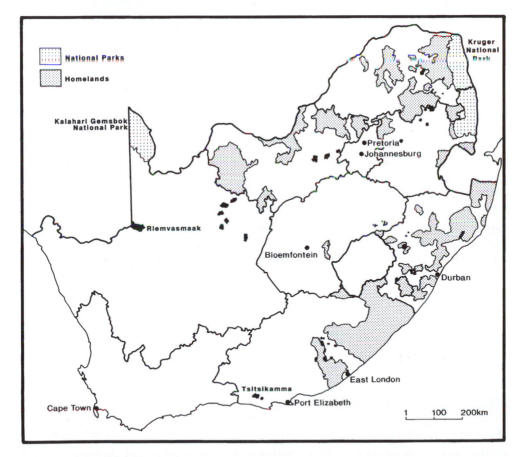

Figure 8.10 Rural land restitution
Source: Modified after A.J. Christopher (1997) *Spatial Aspects of Indigenous Lands and Land Claims in the Anglophone World*, Cambridge: University of Cambridge, Department of Land Economy

forms of rights were also claimed, including those derived from long-term urban tenancy, rural labour tenancy, share-cropping and grazing rights. Many claimants were unable to provide documentation, so necessitating a time-consuming research investigation. The five-year period allocated for completion of the task was highly optimistic, in view of the complexity of the legislation, as only twenty-six claims covering 167,000 hectares had been resolved by the end of 1998 (South Africa 1999: 104). A more flexible approach was adopted in 1999 to expedite the process.

Rural land claims generally fell into three categories depending upon ownership and occupation. First, where the land had been transferred to White farmers, some were willing to sell back to the state for restitution. One of the first such restorations took place in 1994 at Tsitsikamma, where the Mfengu community was returned to their lands (see Figure 3.10, p. 81). Second, where the land had been retained by the state for military purposes, there was resistance from the Department of Defence. Many of the major military bases created during the 1970s and 1980s, including much of the battle school at Lohatla, had been built on land acquired through the dispossession of rural African communities. The restoration of the Riemvasmaak community in the Northern Cape in 1994 indicated that the army could surrender land without sacrificing efficiency (McKenzie 1998). Third, where land had been taken for nature conservation areas, reoccupation was generally not possible. But the agreement returning part of the Kruger National Park to the Maluleke people in 1998 again indicated that nature conservation need not clash with established land rights, when management of a section of the park was offered in restitution. Similarly the restoration of land in the Kalahari Gemsbok National Park to the San population in 1999 marked a significant step in the preservation of a highly endangered culture and language. However, in general restitution claims against the National Parks Board and the South African National Defence Force were more difficult to resolve than those against White farmers.

Parallel with restitution went an African land settlement programme, conducted under the Provision of Certain Land for Settlement Act of 1993 (subsequently retitled the Provision of Land and Assistance Act). A number of districts were identified in each province for pilot reform schemes in order to concentrate resources and assist those relocated (Figure 8.11). Land was usually obtained on the open market through the utilization of the land acquisition grants offered to each family without land or the financial resources to gain access to land. Progress again was slow as the resources of the Department of Land Affairs were limited and several ambitious reform programmes were undertaken at once (South Africa 1999). The official aim of redistributing 30 per cent of the arable land of the country within five years had more modest results, with only an approximate 1 per cent change in the racial balance of rural land holdings. As a result, the glaring disparity, between the 55,000 white farmers possessing some 102 million hectares and the 1.2 million African 'micro-farmers' who share about 17 million hectares of the former homelands, remains as great as ever.

THE POST-APARTHEID CITY

The impact of change in the South African city was evident whether in the major metropolises or smaller towns. The physical structures of the apartheid era were relatively unaffected, but social and economic issues began to dominate new developments, where political issues had been paramount before (Davies 1996; Kotze and Donaldson 1998; Lemon 1995). Most publicity was directed towards the central areas where the twin

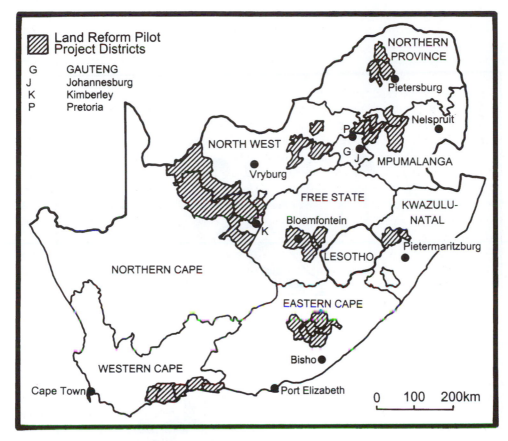

Figure 8.11 Land reform pilot projects
Source: After Department of Land Affairs, *Land Info*, Vol. 1/1996, p. 19

processes of the decline of the Central Business District and the transformation and physical deterioration of the inner residential areas, noted in the late apartheid city, gained greater impetus (Guillaume 1997; Morris 1997). For example, the change in the occupation of central Durban between 1985 and 1996 may be examined by comparing Figures 4.24 (p. 131) and 8.12. The survival of the Warwick Avenue pocket of integration is notable (Maharaj 1999). Initially the integration of the high-density flatlands on the Bay was most evident (Maharaj and Mpungosa 1994). After 1991 the process spread with the reoccupation of residential premises in the Indian Business District and a general influx of people elsewhere into the Central Business District and the inner suburbs. Significantly the urban land restitution programme had not commenced by this date and so the reconstruction of the disrupted multi-racial communities had not occurred. Indeed the inter-community conflicts over future plans for projects in Cato Manor, as in District Six in Cape Town, delayed any rapid redevelopment.

However, at the same time much of the increase in the urban population continued to be accommodated in informal settlements and site-and-service schemes on the periphery of the cities, resulting not only in sprawl but also in a degree of infilling the extensive open spaces

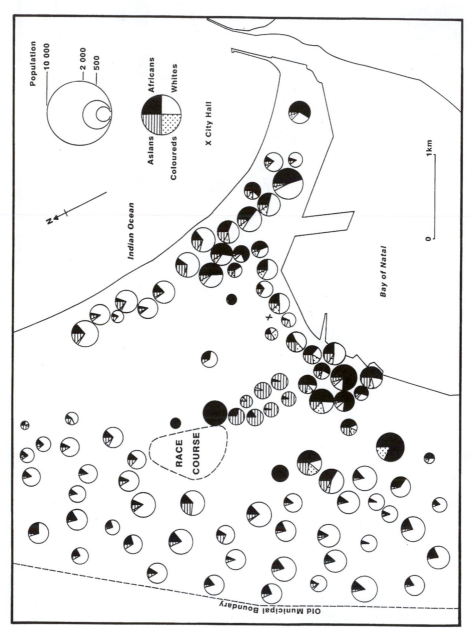

Figure 8.12 Distribution of population in central Durban, 1996
Source: Based on statistics from Statistics South Africa, Pretoria

and buffer strips of the apartheid city plan. The migration of Africans into the small urban centres took place largely within the apartheid spatial framework (Krige 1997). In contrast the majority of homeland towns, more especially those created as resettlement centres, ceased to grow as their *raison d'être* had been removed (Krige 1996).

The wide publicity given to crime, particularly in the city centres, had the result of encouraging the owners of shops and businesses to relocate to suburban centres. Thus many of the city-centre functions of Johannesburg, including the stock exchange, were relocated to Sandton in the course of the 1990s. As the outer suburbs remained essentially segregated on a racial basis, the result was the relocation of the White central business district to a still predominantly White area. Semi-fortified housing complexes, gated roads and walled suburbs with private security patrols were the extreme manifestations of the perception of the threat posed by crime in the former White suburbs (Judin and Vladislavic 1998).

In contrast, the African suburbs began to undergo transformation through the municipal upgrading of services and individual improvements to properties. At the same time the reduction of the high densities in the African suburbs was accompanied by the resettlement of backyard shack dwellers to their own plots in new suburbs financed by land acquisition grants. Land invasions also became a significant aspect of urban development in the 1990s (Gigaba and Maharaj 1996; Saff 1996). The ruthless tactics of bulldozing informal settlements adopted in the apartheid era were not so readily employed after 1990, although the government maintained a highly restrictive policy. Thus the selection of land and the organization of land invasions became important as a critical minimum size was necessary to prevent the authorities from demolishing temporary dwellings. Those invasions adjacent to existing African areas were usually regularized and incorporated into the town plan, as noted in Bloemfontein (Krige 1998). Other land invasions were more likely to be relocated to land deemed more suitable by the local authorities, as noted in Hout Bay in Cape Town (Oelosfse and Dodson 1997).

The changing distribution of population was captured by the 1996 census, which, in line with the official concept of multi-racialism, retained the racial classification question. The basic apartheid four-fold classification was complicated by the introduction of an 'other and unspecified' category, in order to accommodate the plea for recognition by the Griqua and other groups previously classified as Coloured. Some people refused to give a designation and were left unclassified by the enumerators. However, less than 1 per cent of the population was returned in this category, permitting a comparative analysis with earlier enumerations.

The most surprising result of the census was the 12.5 per cent decline in five years in the number of Whites, reflecting the experience of other post-independence African states and emphasizing the exodus to other English-speaking countries such as Australia (Rule 1994). Although not as dramatic as the White flight from Zimbabwe in the 1980s, the emigration offered opportunities to other groups in the property market (Cumming 1993). The urban population of the country reached 21.8 million, representing over half the total, and more than double the 10.3 million enumerated in the previous all South Africa census in 1970. Remarkably, the White component of the urban population had shrunk to only 20 per cent, compared with 31.6 per cent in 1970.

In the specific case of Port Elizabeth, the enumerated population reached 750,000 inhabitants in 1996 (Figure 8.13). In terms of distribution, infilling of open spaces and central city integration are noticeable, compared with the highly segregated situation only five years before. The Africanization of sections of the previously virtually all-White city centre is indicative of the transformation taking place in most towns and cities. On the other

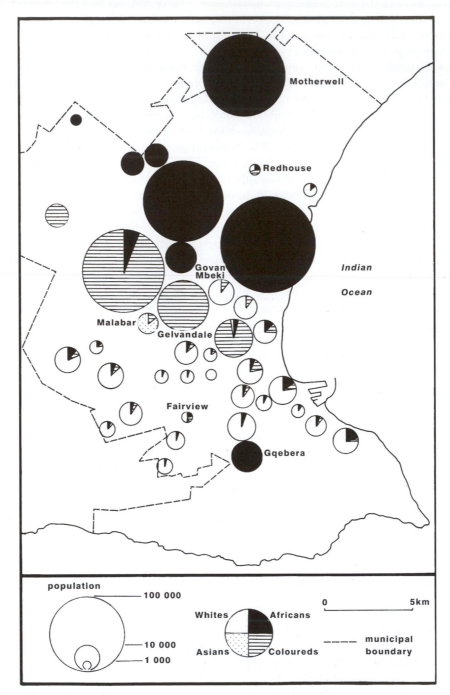

Figure 8.13 Distribution of population, Port Elizabeth, 1996
Source: Based on statistics from Statistics South Africa, Pretoria

hand, the new peripheral settlements remained largely uni-racial and the inherited apartheid framework was hardly breached. The beginning of the rebuilding of the multi-racial suburb of Fairview and the expansion of Walmer Township, renamed Gqebera ('Emerald Hill'), are indicative of these divergent trends.

Urban segregation levels declined throughout the country between 1991 and 1996. The key White index of segregation witnessed widespread, but relatively small, reductions (Figure 8.14). This was mainly a result of members of previously excluded communities returning to take up residence in the former White group areas as the first wave of enfranchised and relatively wealthy 'pioneers' entered the property market. It is notable that small towns remained highly segregated as the general economic stagnation of the rural areas reduced opportunities for change, while the accommodation styles of the flatlands and inner suburbs which had attracted the initial wave of change in the large centres were largely missing in the small towns. Thus, for example, Graaff-Reinet, a town of 34,000 inhabitants, remained highly segregated five years after the repeal of the Group Areas Act (Figure 8.15). Its experience was reflected in most other small towns. Even the former homeland capitals of Umtata and Mmabatho, where urban apartheid had been dismantled at independence, remained remarkably segregated, with the key White index measuring over 80 in 1996, indicative of the inertia of segregation.

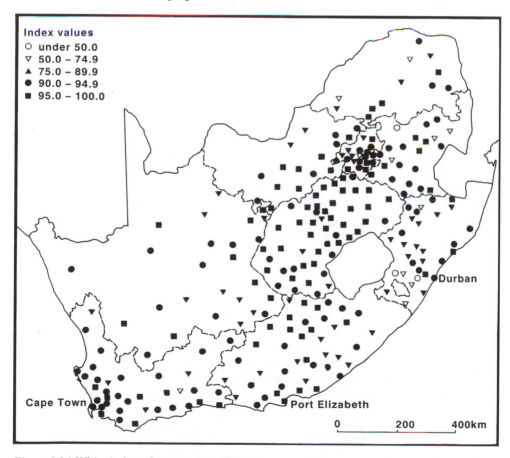

Figure 8.14 White index of segregation, 1996
Source: Based on statistics from Statistics South Africa, Pretoria

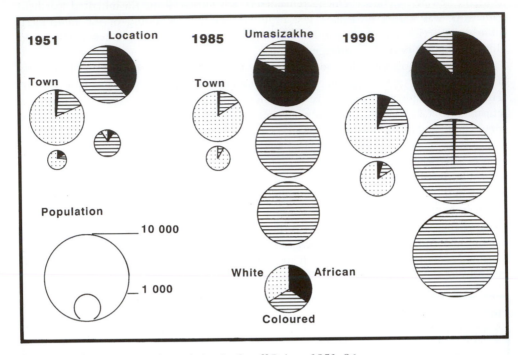

Figure 8.15 Distribution of population in Graaff-Reinet, 1951–96
Source: Based on information then held by the Central Statistical Services (1951 and 1985) and
Statistics South Africa (1996), Pretoria

The post-apartheid city was thus still constrained by the inherited physical framework of
the apartheid city (Figure 8.16). In contrast to the city centres and inner suburbs, the outer
suburbs remained significantly segregated on racial lines, but might more accurately be
described as high density and low-density. This was particularly noticeable for the majority
of Africans who were too poor to move to the former White, low-density, suburbs experi-
encing integration. Yet increasingly the suburbs were perceived as a reflection of economic
status, where race and economic status remained closely, but not exclusively, linked.
Opportunities for significant spatial changes were limited. However, the availability of open
land near city centres and within the low-density suburbs offered the chance to relocate
poorer people in need of housing close to high-status areas. In this manner serviced land
could be provided relatively cheaply as the basic infrastructure of main roads, water and
sewerage were accessible. At the same time pressures to redevelop existing residential areas
to higher densities resulted in the accommodation of greater urban populations with relatively little outward extension of the built-up area, thus making the cities more compact.

From a planning perspective, the ordering of South African cities in some measure passed
from the hands of professional city planners, creating more flexible structures (Mabin 1995;
Mabin and Smit 1997). Indeed community-based civic organizations exerted considerable
influence in determining the siting and form of new housing schemes (Maharaj 1996b). In
certain respects functional segregation decreased as the flight of businesses and shopping
facilities from city centres continued, with clienteles reflecting the adjacent suburbs and so
creating a multi-cellular city, without a distinct unitary core or central business district. In

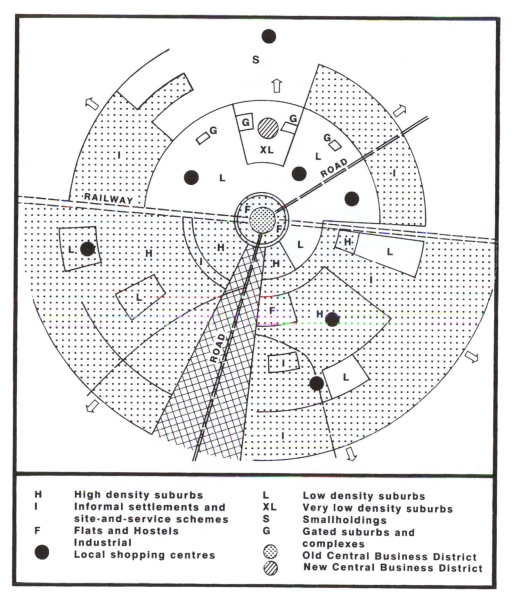

H High density suburbs   **L** Low density suburbs
I Informal settlements and   **XL** Very low density suburbs
site-and-service schemes   **S** Smallholdings
F Flats and Hostels   **G** Gated suburbs and
Industrial   complexes
● Local shopping centres   Old Central Business District
New Central Business District

Figure 8.16 The post-apartheid city?
Source: Compiled by the author

this respect the contemporaneous reordering of Soviet and South-East Asian cities on more overtly free market lines offered some parallels (Dick and Rimmer 1998; Lehmann and Ruble 1997).

DISMANTLING PERSONAL APARTHEID

The repeal of the Reservation of Separate Amenities Act and the removal of the various discriminatory regulations governing personal relations did not immediately give more than nominal equality to all the citizens of South Africa. Park benches and building entrances might no longer be marked 'Whites' and 'Non-Whites', but in many spheres of life little changed. However, in a certain number substantial changes were effected.

In the realm of education the removal of apartheid regulations opened up new opportunities at tertiary level. In certain respects this reflected the desegregation of residential areas, in that the existing African institutions did not become integrated, but other establishments enrolled African students in considerable numbers, as the total of university students almost doubled between 1985 and 1995. In the first phase the University of the Western Cape and the University of Durban-Westville lost their ethnic Coloured and Indian character as African students enrolled at the two institutions which were militantly at the forefront of higher educational transformation.

Thereafter the White residential universities experienced substantial increases in the registrations of previously excluded students. By 1997 the numbers of both Indian and African students exceeded the number of White students at the University of Natal, while at the remainder of the previously White institutions, Whites formed little more than half the total student body (Figure 8.17). Only at the University of Stellenbosch, with its elite Afrikaans tradition, did White students continue to constitute more than 80 per cent of the student body, although postgraduate courses were offered in English. The other Afrikaans-medium universities adopted parallel classes in English at both undergraduate and postgraduate levels as a means of attracting African students. Most successful was the Rand Afrikaans University in Johannesburg, where by 1995 White students accounted for only half the total. Within the rapidly expanded tertiary technical education sphere, the transformation was even more dramatic with only the Cape Technikon retaining a White student majority by 1997.

At school level the unification of the seventeen departments of education into a national system was problematical, with the massive inherited disparities in staffing and facilities. The poorly resourced rural homeland schools required substantial investments, which were only slowly made available. It was in the urban areas that the most notable changes took place, with the breaking down of the previous apartheid divisions. Initially the governing bodies and parents of many formerly White schools sought to exclude children of other groups, particularly African children, through the imposition of entry language tests and the levying of fees to pay for extra facilities and additional staff. However, in many schools, the declining numbers of White pupils softened this approach in an attempt to prevent the redeployment of teachers to understaffed schools. Because of the perceived better facilities and better-trained staff in the White schools, the demands for entry were overwhelming. Schools therefore were made to reflect the emergent, colour-blind economic class structure more closely than most spheres of post-apartheid society. Schooling consequently came to involve substantial commuting of pupils into the former White areas, where the integrated schools were situated (Figure 8.18).[1] Again because of the medium of instruction, the

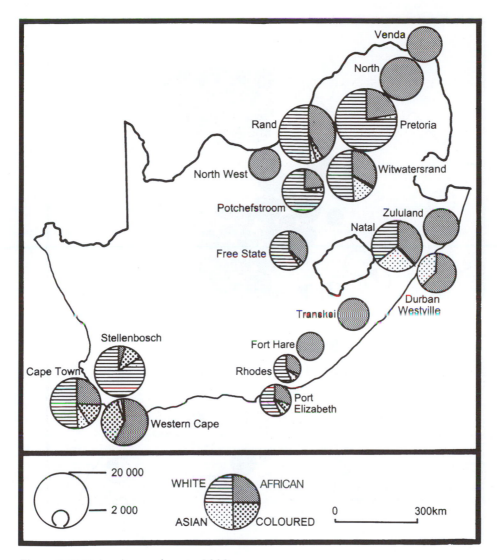

Figure 8.17 University enrolments, 1998
Source: Compiled from statistics in *South Africa Survey 1997/8*, Johannesburg: South African Institute of Race Relations

English-language schools, whether formerly White, Coloured or Asian, were more open to practical and successful integration than Afrikaans-medium schools.

In another context the removal of restraints on individual freedom was extended to dismantling the complex legal restrictions on gambling. The National Lottery and Gambling Act of 1996 ended the homelands' monopoly on casinos, providing for the granting of up to forty licences throughout the country (Figure 8.19). The new enterprises were concentrated in the main cities and many of the homeland resorts were deprived of their *raison d'être* and were closed. Thus six casinos were opened in Gauteng, including Caesar's adjacent to Johannesburg International Airport. In contrast, all but one of the homeland

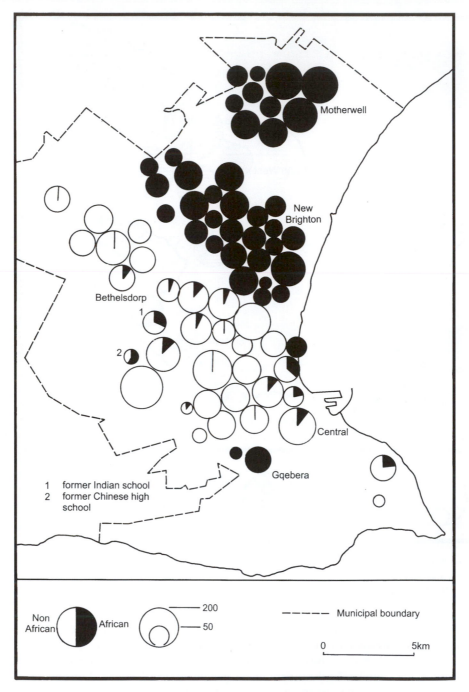

Motherwell

New
Brighton

Bethelsdorp

1

2

Central

1   former Indian school
2   former Chinese high
    school

Gqebera

Non
African   African

200

50

Municipal boundary

0   5km

Figure 8.18 High school enrolments in Port Elizabeth, 1998
Source: Compiled from information in *Eastern Province Herald*, 8 January 1999, Special
Supplement: Eastern Cape Matric Results 1998

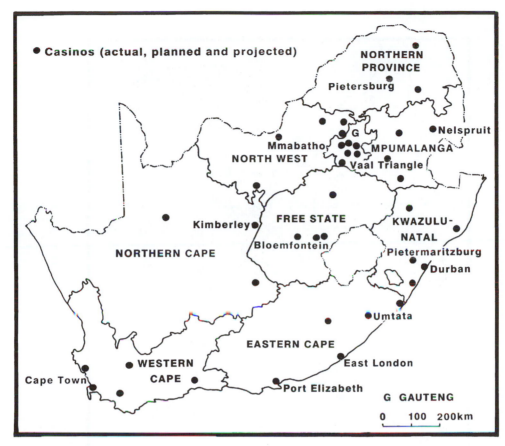

Figure 8.19 Casinos in South Africa, 1999
Source: Compiled from information supplied by the provincial Gambling Boards

resorts in the Eastern Cape were closed in favour of new resorts in Port Elizabeth and East London. Few people in the country were now more than 200 kilometres from gambling facilities.

ENDING INTERNATIONAL ISOLATION

The independence of Namibia and the beginning of the internal reform process in 1990 offered South Africa the entry into the international community which it had previously been denied. The National Party government attempted to overcome economic sanctions and launched a diplomatic offensive to gain official recognition. The main region of success was eastern Europe, where former communist governments were also engaged in major structural reform programmes and had little commitment to the anti-apartheid struggles of the past (Figure 8.20). The practical need for South Africa to establish embassies in Slovakia or Bulgaria might be questionable but the political benefits of doing so were regarded as outweighing any financial loss.

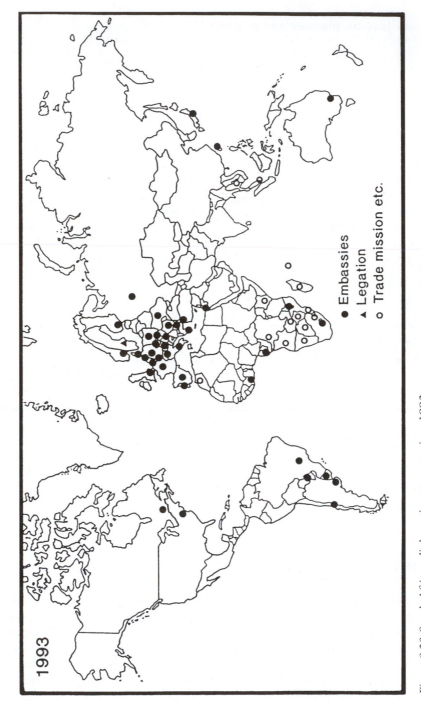

Figure 8.20 South African diplomatic representation, 1993
Source: Compiled from South Africa (1994) *Department of Foreign Affairs List*, Pretoria: Department of Foreign Affairs

Relations with African countries also underwent a remarkable change. Three additional countries exchanged ambassadors with the National Party government. Gabon and the Ivory Coast had been particularly supportive of the government's reform initiatives. Lesotho was totally dependent upon South Africa and considered the move expedient. In addition, a number of countries exchanged trade representatives or other functionaries who acted effectively as diplomatic representatives, but did not break the diplomatic sanctions still kept in place by the Organization of African Unity. As a result, between 1990 and 1993 the government had been able to increase the number of diplomatic missions abroad from twenty-three to thirty-nine.

Diplomatic sanctions were symbolically ended with the inauguration of President Mandela on 10 May 1994. The governments of over 130 countries were represented at the ceremony in front of the Union Buildings in Pretoria in one of the largest diplomatic gatherings since the Second World War. Leaders as diverse as Vice-President Al Gore of the United States and President Fidel Castro of Cuba headed the guest list. The country was readmitted to the Commonwealth, and joined the Organization of African Unity, the Southern African Development Community and the Non-Aligned Movement, and took up its seat in the United Nations once more.

There followed a major expansion (to seventy-one) in the number of South African ambassadors and high commissioners resident around the world by 1999 (Figure 8.21). The majority of the new missions were in Africa, although Francophone West Africa remained comparatively under-represented. It may be noted that Europe still narrowly outnumbered Africa in the number of resident ambassadors. In addition new embassies were established in the major Asian countries, including Iran and Indonesia, where sanctions had been strictly enforced before 1994. A symbolic embassy was also established in Palestine, as an indicator of a changed national world-view. A major political transfer took place in 1998 with the exchange of ambassadors with the People's Republic of China and the downgrading of relations with the Republic of China (Taiwan). In the Americas, Cuba and Mexico hosted resident ambassadors, while the embassies were closed in Paraguay and Uruguay, which had been the scenes of the diplomatic offensive of the 1970s, as were several in eastern Europe dating from the early 1990s.

The ending of South Africa's isolation was no better illustrated than by the invitations extended to President Mandela to visit other countries.[2] National leaders wished to be seen and photographed with the South African President, who assumed a high moral, almost iconic, status. It was testimony to this status, and his willingness to travel and mediate, that in 1999 he was able to facilitate a settlement of the Anglo-American dispute with Libya over the sabotage of a Pan Am airliner over the town of Lockerbie in Scotland in 1988. In the course of his five-year presidency some fifty-seven countries were included in his itinerary, either for official and state visits or to attend national celebrations or meetings of international organizations (Figure 8.22). The spread of these countries was noticeably African (twenty-two) in orientation, followed by Asia (sixteen) and Europe (thirteen). Indeed the first five countries which he visited were African and half the presidential visits were to other African countries as some southern African countries extended several successful invitations, demonstrating their continuing solidarity with the African National Congress. In terms of time, President Mandela spent a total of seven months of his presidency outside the country, with two months apiece in Africa and Asia.

It was indicative of the changed priorities of the new government that it was admitted to the renamed Southern African Development Community (SADC), an organization originally formed to counter South African influence in the region (see p. 187). Inevitably, South

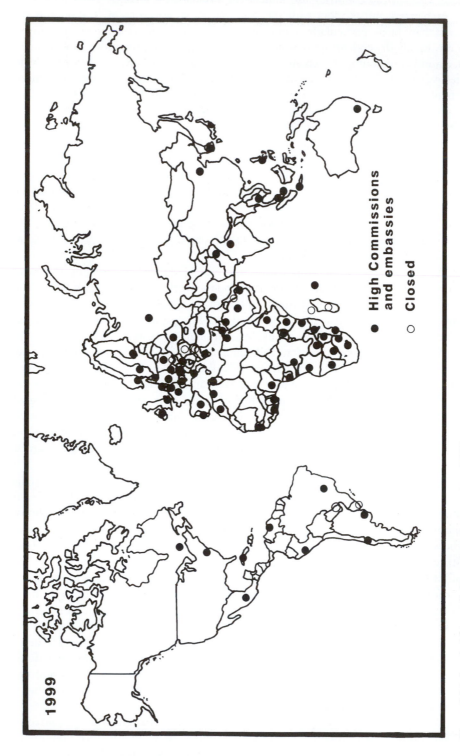

Figure 8.21 South African diplomatic representation, 1999
Source: Compiled from South Africa (1999) *Department of Foreign Affairs List*, Pretoria: Department of Foreign Affairs

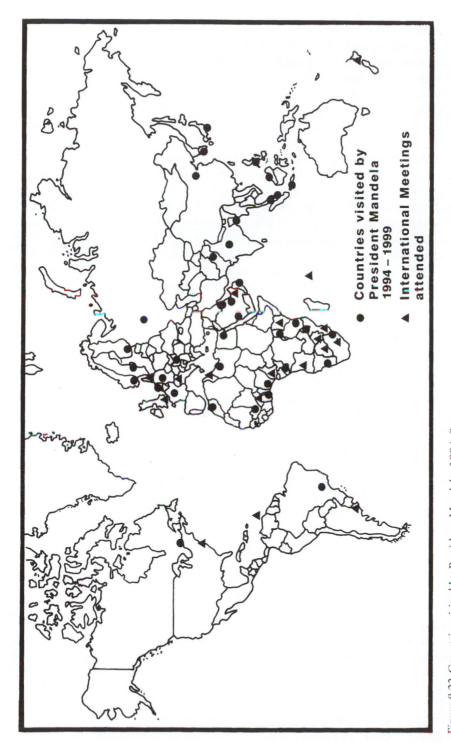

Figure 8.22 Countries visited by President Mandela, 1994–9
Source: Compiled from information supplied by the Subdirectorate: State Visits, Department of Foreign Affairs, Pretoria

Africa dominated the grouping in economic terms, but was less significant in military and political terms (Gibb 1998). However, the moves to expand the organization and to give it a more integrative role in the region became a South African priority (Figure 8.23). Mauritius and the Seychelles were admitted to the SADC. The most substantial change was the admission of the former Zaire (inappropriately renamed the Democratic Republic of Congo) after the overthrow of President Mobutu Sese Seko in 1997. Thus the last state linked to the South African Railways standard gauge network was included in the Community. The subsequent Congolese civil war divided the organization, as its Inter-State Defence and Security Committee, under the leadership of Zimbabwe, was involved in the military intervention to support the new government. Mindful of the destabilization policies of its predecessor, the Government of National Unity attempted to avoid military intervention in the region, although the operation in Lesotho in 1998 demonstrated the potential for a more aggressive foreign policy. South Africa also chose to remain outside the Common Market for Eastern and Southern Africa (COMESA), which because of its greater extent, from Egypt and Ethiopia to Lesotho and Madagascar, was not so amenable to the country's demands.

Economic sanctions against South Africa had been one of the most effective means of forcing change and reaching a negotiated settlement within the country. After 1990 the National Party government therefore considered that once the negotiation process had begun, sanctions should be lifted. However, the African National Congress supported their continuation until the process of change was demonstrably 'irreversible'. It was a measure of the basic mistrust between the two organizations that no agreement on the issue could be reached. The American Comprehensive Anti-Apartheid Act was repealed in 1991, when the two governments agreed that the specific conditions laid down in the Act had been met. Other sanctions were removed in a piecemeal fashion, and the United Nations embargoes on oil and arms sales were not lifted until the new government was installed in 1994.

The spatial dimensions of the imposition of sanctions concerning international trade and the national airline were illustrated in Chapter 7. The dismantling of both in the course of the 1990s may be taken as illustrations of a return to a non-politicized 'normality'. In the case of trade sanctions, most measures were only repealed in the years 1992 to 1994. Thus the 1993 pattern of South African exports still showed the effects of sanctions, although the remarkably high percentage (37 per cent) of goods not assigned a specific destination may indicate sanctions breaking (Figure 8.24).

It was in the period immediately after the installation of the Government of National Unity that the effects of the lifting of sanctions were recognized. Between 1993 and 1998 the Rand value of exports nearly doubled, but this merely reflected the parallel decline in the dollar value of the currency (see Figure 0.3, p. 7). The geographically unassigned category declined to 26 per cent (mostly precious metals and stones bound for markets in Europe). Shifts in trade patterns were significant. The European share of trade, which had declined in the early 1990s, remained constant. It was the return of South African exporters to Africa which was the most marked change, with the continent's share rising to 14 per cent, a level not witnessed since the 1970s. Remarkable shifts took place within the broad continental regions, with states such as India and Indonesia becoming significant importers of South African goods. On the other hand, exports to Taiwan still exceeded those to the People's Republic of China by a factor of 3.5, which remained virtually unaltered between 1993 and 1998. The serious impact of the Asian economic recession was evident in the decline in exports to that continent from 21.1 per cent in 1997 to 16.6 per cent in 1998.

The lifting of sanctions on international flights was more immediate. In 1990 foreign carriers returned to South Africa in significant numbers. The ban on overflying Africa

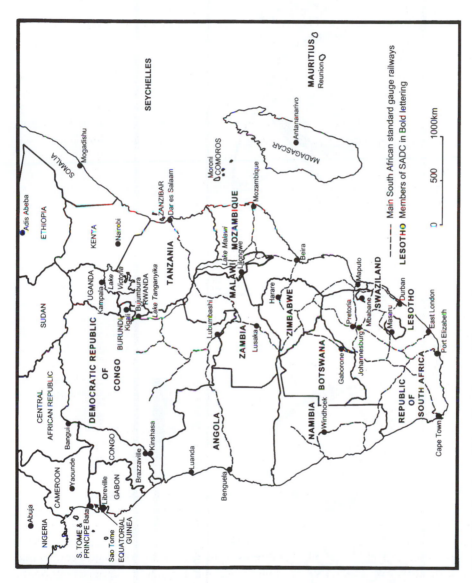

Figure 8.23 Southern African Development Community
Source: Compiled by the author from various sources

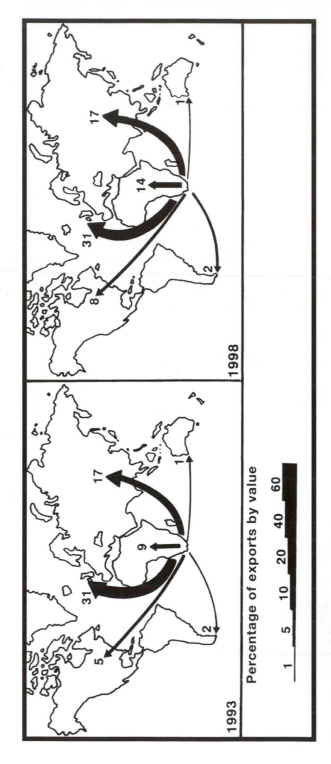

Figure 8.24 South African exports, 1993–8
Source: Based on statistics in the 1993 and 1998 *Monthly Abstract of Trade Statistics*, Pretoria: Government Printer

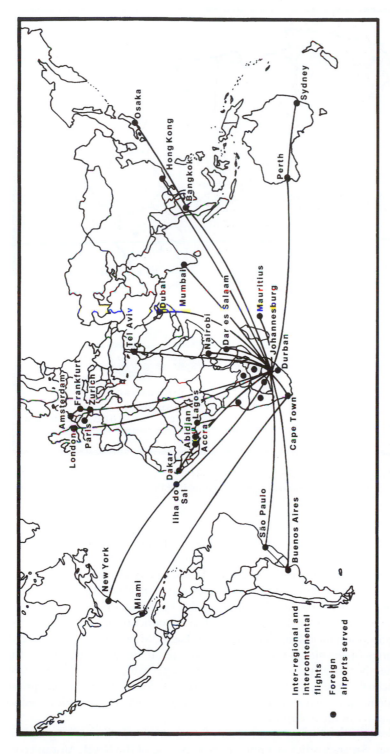

Figure 8.25 South African Airways intercontinental flights, 1999
Source: Based on South African Airways 1999 Timetable

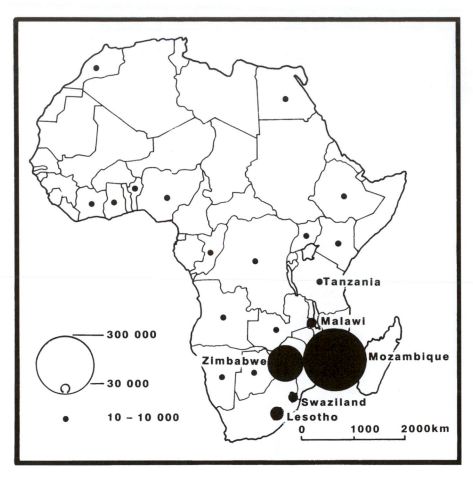

Figure 8.26 'Illegal aliens' in South Africa
Source: Based on statistics in *Hansard*, National Council of Provinces, 14 May 1998, cols. 332–40

imposed on South African Airways was also lifted, enabling the more direct routes over Africa to Europe and Tel Aviv to be used. In the United States the lifting of sanctions allowed the reopening of the route to North America in 1991, while the direct link with Australia was revived at the same time. The management of South African Airways now found itself in the remarkable position of being able to choose destinations on the basis of profitability, rather than political expedience. The introduction of new services to African and Asian destinations, including flights to Dakar, Lagos, Mumbai and Osaka, stretched the limited resources of the airline (Figure 8.25). As a result some existing routes had to be cancelled. Thus the direct air links with Lisbon and Taipei, which had been considered to be politically essential in the period of sanctions, were abandoned. Furthermore, privatization and cooperation agreements, notably with airlines elsewhere in Africa, reduced the 'national' character of the enterprise.

South Africa also became the destination of a remarkable inflow of migrants from the countries of sub-Saharan Africa as a result of the opening of the South African economy to the forces of globalization (Stalker 1999). The 'illegal aliens' controversy with the attendant xenophobia which it sparked in South Africa in the late 1990s was a measure of the attractive nature of the major economic powerhouse on the continent (Crush 1999). The free migration differed from the controlled mine recruitment, which itself was undergoing reassessment (Crush and James 1995). Estimates of the number of people involved ranged substantially from 500,000 by the Department of Home Affairs to 2.5–4.1 million by the Human Sciences Research Council. These figures included a substantial number (over 300,000) of Mozambican refugees in Mpumalanga who had fled from the devastation caused by the civil war in their own country. An amnesty offered by the Department of Home Affairs in 1996 only netted 200,000 applicants, half of whom were subsequently denied permanent residence status and deported.

Deportations in the two years 1996 and 1997 amounted to over 350,000 people. Mozambicans constituted some 85 per cent of the total, followed by Zimbabweans, Basothos and Malawians (Figure 8.26). The only other countries with significant numbers of deportees were Swaziland and Tanzania. Studies suggested that the migrants brought organizational and entrepreneurial skills which were highly useful to the South African economy in the period of reconstruction which began in the 1990s (Rogerson 1997). Furthermore, the majority of the migrants were temporary and intended to return home once a set of predetermined economic objectives had been achieved, following the pattern of many earlier migrants.

9 Legacy of apartheid

The previous chapter concentrated upon the dismantling of apartheid. It recorded that in certain spheres of activity apartheid has been effectively deconstructed. The new constitution, together with the new governmental structures, is essentially non-racial and administered on a colour-blind basis. Following this achievement, significant efforts have been made to promote a feeling of united nationhood in which all are active participants. Archbishop Desmond Tutu's concept of the 'rainbow nation' was significant as an attempt to blend together the disparate ethnic groups promoted separately by the previous government, and give them a sense of common purpose, while they retained their own individuality (Ramutsindela 1997). This concept of diversity was complemented by the promotion of an awareness of the common African heritage of the citizenry as part of the realization of an African Renaissance espoused by President Thabo Mbeki (1998). Such a national concept has evoked varied responses in other African states which recognize the diversity of identities and relationships to the government and its leadership (Mamdani 1996). It also reflects the absence in South Africa of 'an ethnic core around whose values (language, history, mythology) a nation could be constructed' (Simpson 1993: 19). Tokyo Sexwale, first premier of Gauteng, appealed for a more inclusive and unified view of the nation: 'If blacks get hurt, I get hurt. If whites get hurt, that's my wife, and if you harm coloured people, you're looking for my children. Your unity embodies who I am' (*Sunday Times*, 27 June 1999: 19).

Some of the other distortions of the apartheid era are in the process of being undone, creating new landscapes through land reform and urban reconstruction and development. At another level a sense of change and of a new era can be engendered simply through the renaming of places. Thus the principal military base situated in the hills outside Pretoria was renamed Thaba Tshwane (after the illustrious eighteenth-century African leader). It had been named Voortrekkerhoogte (after the Afrikaner pioneers of the nineteenth century) during the apartheid era and before that Roberts Heights (after the British Commander-in-Chief in the Anglo-Boer War), so reflecting the dominant political groups of the time. Its adjacent municipality underwent similar renaming as Centurion (previously Verwoerdburg, and before that Lyttelton), as did the province of Gauteng (previously Transvaal) in which it is sited. At the same time the reversion of Triomf to its previous name, Sophiatown, was highly symbolic in the sense of 'undoing' the apartheid past. The opportunities for the official manipulation of place names as a means of projecting changes of identity remain to be fully exploited.

However, the legacy of the colonial and apartheid eras remain in many other spheres of life. The African and White experiences of the country remain different and unequal, whether in inherited disparities of income and education, or in the suburb or rural area in which they live. Formal segregation may have been removed but many aspects of the geography of apartheid remain in place. It is proposed to examine two of these, namely education and income, as illustrations of the wider range of inequalities. The extreme poverty of the African rural areas must also not be forgotten: it is hidden, in comparison with that of the slums in the cities which suffer the more publicized 'urbanization of poverty'.

INCOME

It was the marked inequalities in income that prompted the then Deputy President, Thabo Mbeki, to observe in an address to parliament in 1998 that:

> A major component of the issue of reconciliation and nation-building is defined by, and derives from, the material conditions in our society which have divided our country into two nations, the one black and the other white. We therefore make so bold as to say that South Africa is a country of two nations. One of these nations is white, relatively prosperous, regardless of gender or geographic dispersal. . . . The second and larger nation of South Africa is black and poor, with the worst-affected being women in the rural areas, the black rural population in general and the disabled. This nation lives under conditions of grossly underdeveloped economic, physical, education, communication and other infrastructure.
>
> (*Hansard* 1998: col. 3378)

In 1995, the October Household Survey found that whereas 5.5 per cent of Whites were unemployed, the African percentage was 36.9 (South African Survey 1998). Furthermore, unemployment levels were regionally skewed: 41 per cent of the economically active population in the Eastern Cape and Northern Province were without work, compared with 18.6 per cent in the Western Cape.

When converted into incomes the inequalities become extremely bleak, with the African annual average household income under a quarter of that of White households (South African Survey 1998). Furthermore, the average household incomes vary substantially on a provincial basis, with all groups in Gauteng substantially better off than elsewhere (Figure 9.1). Programmes of income equalization through the promotion of the Growth, Employment and Redistribution policy instituted in 1997, and affirmative action in staff appointments, under the Employment Equity Act of 1998, remain far from completion.

EDUCATION

One of the most effective ways to eliminate these economic inequalities is through education. Indeed the Government of National Unity devoted a significant proportion of the national budget to this end. However, the educational profiles of the various population groups remain remarkably different. Thus the 1996 census revealed that whereas 64.8 per cent of the adult White population had completed secondary education, only 15.1 per cent of adult Africans had done so (South Africa 1998b). Conversely, whereas 24.3 per cent of the African adult population had received no formal education, the White percentage was only 1.2. When these racial disparities are merged at provincial level the relative advantages of Gauteng and the Western Cape are evident, when compared with the educational backlogs in the Northern Province and Eastern Cape (Figure 9.2).

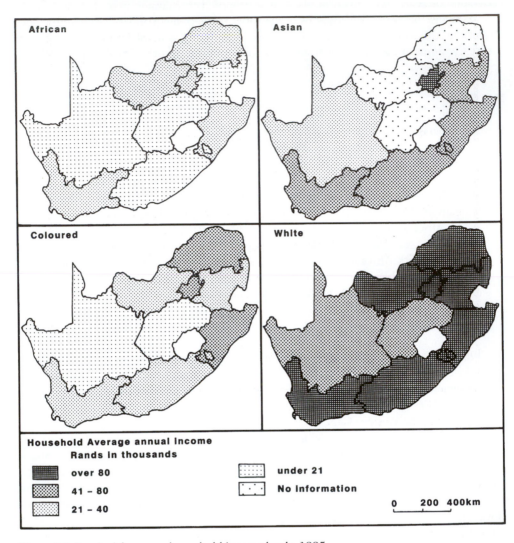

Figure 9.1 Provincial average household income levels, 1995
Source: Based on statistics in *South Africa Survey 1997/98*, Johannesburg: South African Institute of Race Relations

Against this background, the levels of educational attainment remain highly unequal, perpetuating existing inequalities. The effects of a lack of resources, overcrowding and frequently the lack of motivation on the part of both the staff and the pupils in many African schools have been impossible to overcome on a short-term basis, prompting the Minister of Education to describe the system as being in 'crisis' in 1999 (*Eastern Province Herald*, 28 July 1999). Thus the provincial range of results for the final matriculation examinations written at the end of 1998 was substantial and indicative of the continuing apartheid legacy (Figure 9.3). The educational problems of the former homelands are particularly evident in the low continuing pass rate in the Northern Province, where 90 per cent of the population lived in the former Gazankulu, Lebowa and Venda before

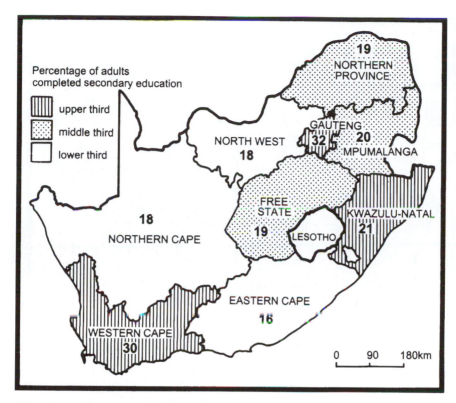

Figure 9.2 Provincial education levels, 1996
Source: Based on statistics in South Africa (1998) *Population Census 1996*, Pretoria: Statistics South Africa

1994. It should be noted that the 1998 Mpumalanga matriculation results were subsequently found to be inflated by approximately 20 per cent as a result of a concerted fraud perpetrated in the education department (*Sunday Times*, 2 May 1999). This was indicative of the high levels of corruption experienced in several sectors of the provincial administration.

These generalized figures hide the vast differences between the legacies of the different education departments. When the results achieved in one city, Port Elizabeth, are compared at individual school level this legacy becomes apparent (*Eastern Province Herald*, 8 January 1999). The acute academic poverty of the majority of, but not all, schools in the former African areas is immediately evident, dependent as school achievement is upon the dedication and effort of pupils, teachers and parents (Figure 9.4). The results become even more stark when a comparison of the rates of attainment of matriculation exemption levels for admission to university is made.[1] Schools achieved results ranging from zero to 92 per cent. The state, formerly White, Collegiate High School for Girls and the Grey High School for Boys, together with the private Theodore Herzl High School, attained results markedly better than all others. Significantly, one-sixth of all university exemption level passes gained by African pupils in the city were gained by pupils in formerly all-White schools, where individual levels of attainment reflected those of the schools, not the racial origins of the

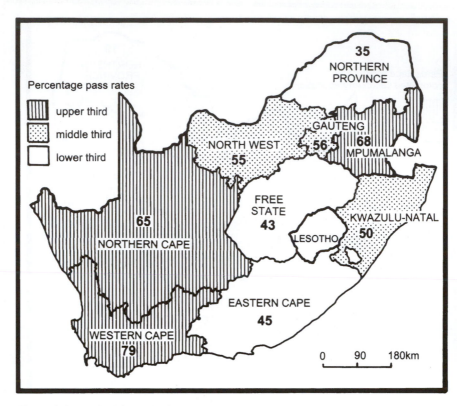

Figure 9.3 Provincial matriculation pass rates, 1998
Source: Based on statistics in *Eastern Province Herald*, 8–12 January 1999

pupils. However, with such poor results only a small proportion of African pupils will be able to achieve the academic and technical qualifications generally deemed necessary to raise income levels appreciably in the highly competitive global working environment in which South Africa will be placed in the new millennium.

TRANSFORMATION

In the next twenty years South Africa will become an increasingly urbanized society, and the growth in the numbers of the population will have to be accommodated within the cities. However, South African cities are the most enduring monuments to apartheid. The apartheid city was built as 'the domain of the White man' and the disparate levels of physical amenities and services provided to the various racially defined sectors emphasized this domination. Indeed, the physical structures created in the apartheid era, whether the grandiose government buildings, churches and monuments on the one hand, or uniform township houses on the other, are likely to be its most enduring legacy. The African townships were considered to be no more than appendages to the White towns and were built and serviced to maintain basic health requirements at minimal cost, so as not to incur expenses which might have to be borne by the White ratepayers.

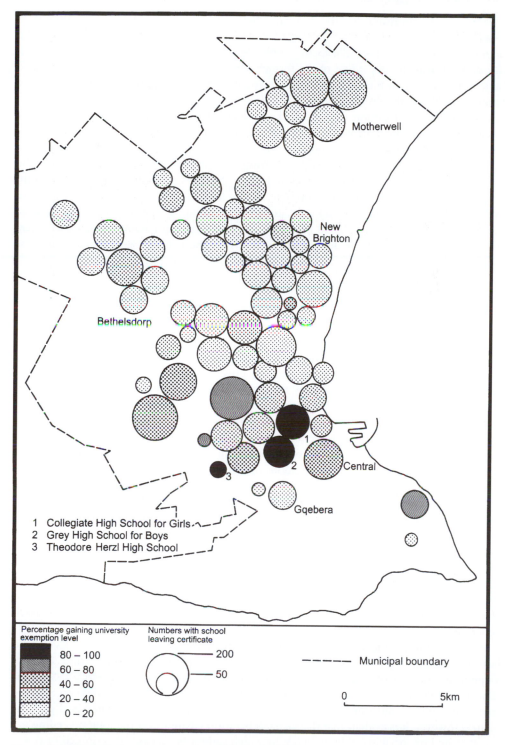

Figure 9.4 Matriculation pass rates, Port Elizabeth, 1998
Source: Compiled from information in *Eastern Province Herald*, 8 January 1999, Special Supplement: Eastern Cape Matric Results 1998

It is scarcely surprising that the disparities between the former African and other areas remain as one of the most noticeable legacies of the urban areas where the majority of the population now live (Port Elizabeth 1999). Since 1994 substantial efforts have been directed towards the upgrading of the former African townships with the construction of basic infrastructure in the form of water, electricity and sewerage provision. In view of the limited funds available, some aspects of infrastructural improvement have been assigned a lower priority and therefore remain markedly inadequate. At the same time the spatial integration programmes aimed at disaggregating the sectoral structure of the apartheid city appear to be attainable, given the political will. The City Integration Programme, launched in 1999, aims to locate low-cost housing throughout the city, thereby better utilizing the existing infrastructure and overcoming the legacy of hypersegregation. In Port Elizabeth, in recognition of the anticipated growth of population, the neighbouring towns are included within the planning scheme, aimed at creating a new metropolitan area of 2.6 million souls by the year 2020 (Figure 9.5). The doubling of the built-up area offers the opportunity to build for integration as surely as the previous government built for segregation.

SUMMATION

The policy of apartheid was a product of the late colonial era and was sustained internationally by the rivalries of the Cold War. It was one of the last of the twentieth-century social engineering projects driven by a strong government. As such, apartheid was conceived and administered as an ideology for the total organization of South African society for the exclusive benefit of the White segment of the population. The policy was implemented and enforced with inflexibility and brutality for more than forty years. In that time the National Party government achieved a high degree of success in promoting personal, urban and state apartheid, but at unsustainable costs. The escalating monetary, military and social burdens placed upon the White population ultimately led to its abandonment.

The physical inheritance of the apartheid era will survive for a very long time. Apartheid social engineers were part of a post-Second World War global movement of professional planners seeking to construct new and improved living conditions for a 'better' society. The physical constructions of the era were substantial and effectively permanent. The fabric of the apartheid cities, the homeland settlement patterns and the infrastructure can be adapted but not erased. Just as colonial inequalities were perpetuated in post-colonial Africa and Asia, so there appears to be little likelihood of their speedy removal in South Africa. The physical inheritance from the apartheid era will be a massive legacy to be overcome by future generations.

In personal terms the Truth and Reconciliation Commission, in seeking to alleviate the bitterness and hurt of the past, was vital for the future cohesion of the country and achieved a significant measure of success in overcoming the apartheid legacy (South Africa 1998c). In similar vein President Mandela and the Government of National Unity sought to ease the transitional phase and gain broad multi-racial support for the new African-based government. However, it is evident that the spatial separation created over the long preceding era was matched by social isolation, with results which cannot be so easily determined or eradicated. The phenomenon of hypersegregation noted elsewhere is reflected in deeper personal alienation which legal enactments cannot undo, and which require individual responses. As a

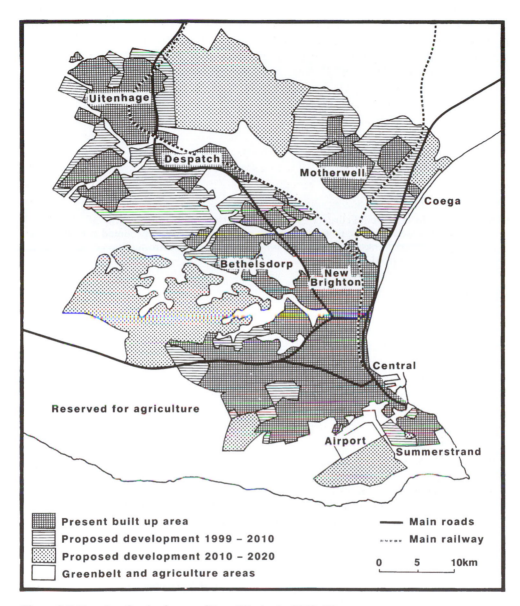

Figure 9.5 Planning for the future of Port Elizabeth, 2000–20
Source: Modified after Port Elizabeth (1999) *The First Comprehensive Urban Plan*, Port Elizabeth: Port Elizabeth Municipality

result, two separate and unequal nations still exist in South Africa and merging them into a united nation is one of the significant tasks confronting both the government and the citizenry at President Mbeki's 'dawning of the dawn' in the new millennium. Only with the success of reconciliation and recognition that the two nations are no longer racially defined will it be possible to pronounce the end of apartheid in South Africa.

Notes

Introduction

1 The Rand was introduced in February 1961, replacing the pound (£) at a parity of 10 shillings (50p). The introduction of the new currency was part of the programme aimed at eliminating vestiges of the imperial links with Great Britain, necessitating the introduction of new coinage without the Queen's head.

1 Before apartheid

1 The coverage in Figure 1.23 may be incomplete, but it is designed to present as comprehensive a picture as possible. In 1911 many locations were informal arrangements with no governmental backing. However, they were *de facto* entities so far as the census enumerators were concerned and this source is likely to be more comprehensive than the *de jure* position.

2 See Deeds Office, Pietermaritzburg, Deed of Transfer 27/1921, Durban.

3 See Deeds Office, Cape Town, Deed of Transfer 9672/1919, Fish Hoek.

4 In Figures 1.27–1.30, indices of segregation are calculated according to the following formula:

$$I_x = \frac{0.5\Sigma|x_i - z_i|}{1 - X/Z}$$

Where: I_x is the Index of Segregation of population X, X represents the total of subgroup X in the city, Z represents the total population of the city, x_i represents the percentage of the X population in the ith area, and Z_i represents the percentage of the total population in the ith area. The resultant index is expressed on a scale from zero (complete integration) to 100 (complete segregation) (Duncan and Duncan 1955).

2 Administering apartheid

1 The Location quotient (LQ) compares the percentage of White male Afrikaners who are members of the Afrikaner Broederbond in district i with the percentage in the country as a whole. The quotient is calculated as follows:

$$LQ_i = \frac{\text{Percentage in Broederbond in District i}}{\text{Percentage in Broederbond in South Africa}}$$

If LQ = 1.0, then the district has the same percentage as South Africa as a whole.
If LQ > 1.0, then the Broederbond is over-represented in the district.
If LQ < 1.0, then the Broederbond is under-represented in the district.

7 International response to apartheid

1 The White index of segregation in Windhoek in 1951 was 74, slightly higher than Pretoria (calculated from the records of the National Archives of Namibia, file SWAA 271).

8 Dismantling apartheid

1 Racial classification has always been regarded as a highly sensitive issue in South Africa. As a result statistics are frequently not available for the various population groups in the post-1994 era. For an assessment of the extent of school integration, the comment by the erstwhile Rector of the Grey High School for Boys in Port Elizabeth is apt: 'when I look as the school, I do not see White boys and Black boys, I see only Grey boys'. However, for the compilation of Figure 8.17, the 1998 matriculation results published in the *Eastern Province Herald* (8 January 1999) have been analysed. African matriculants could be identified through their Xhosa Christian names. Identification on the basis of surnames was unsatisfactory, and no means of separating Coloured and White matriculants was possible as they understandably had similar Christian and surnames. The comment on African matriculation rates attached to Figure 9.4 was based on data derived in the same manner.
2 Department of Foreign Affairs, Subdirectorate: State Visits and Ceremonial: Visits Handled 1994–1999.

9 Legacy of apartheid

1 The 1999 results were even more disturbing. In Port Elizabeth some 9,488 pupils wrote their school leaving (Grade 12) examination. Some 50.5 per cent of those in the former White schools gained a matriculation exemption pass entitling them to enter university. This contrasts with 13.0 per cent of those in the former Coloured schools and 4.7 per cent in the former African schools. The former White schools thus produced 65 per cent of the city's matriculants while accommodating only 20 per cent of those writing the examination.

Bibliography

Abu-Lughod, J.L. (1980) *Rabat: Urban Apartheid in Morocco*, Princeton, NJ: Princeton University Press.

Adam, H. (1994) 'Nationalism, nation building and non-racialism', in N. Rhoodie and I. Liebenberg (eds) *Democratic Nation-building*, Pretoria: Human Sciences Research Council.

Adam, H. and Giliomee, H. (1979) *Ethnic Power Mobilized: Can South Africa Change?*, New Haven, CT: Yale University Press.

Adam, H. and Moodley, K.A. (1992) 'Political violence, "tribalism", and Inkatha', *South African Journal of African Affairs* 7: 21–32.

—— (1993) *The Opening of the Apartheid Mind: Options for the New South Africa*, Berkeley: University of California Press.

African National Congress (1994) *The Reconstruction and Development Programme: A Policy Framework*, Johannesburg: Umanyano.

Alden, C. (1996) *Apartheid's Last Stand: The Rise and Fall of the South African Security State*, Basingstoke: Macmillan.

Bahr, J. and Jurgens, U. (1991) 'Free trading areas in South African cities', *Geography* 76: 254–9.

Barnard, W.S. (1982) 'Die geografie van 'n revolusionere oorlog: SWAPO in Suidwes-Afrika', *South African Geographer* 10: 157–74.

Barrier, N.G. (1981) *The Census in British India*, New Delhi: Manohav Publishers.

Bell, T. (1973) *Industrial Decentralisation in South Africa*, Cape Town: Oxford University Press.

Berat, L. (1990) *Walvis Bay: Decolonization and International Law*, New Haven, CT: Yale University Press.

Bernstein, A. and McCarthy, J. (1990) *Opening the Cities*, Durban: Indicator Project.

Best, A.C.G. and Young, B.S. (1972) 'Capitals of the homelands', *Journal for Geography* 3: 1043–55.

Blenck, J. and von der Ropp, K. (1977) 'Republic of South Africa: is partition a solution?', *South African Journal of African Affairs* 7: 21–32.

Bond, P. and Khosa, M. (1999) *An RDP Policy Audit*, Pretoria: HSRC Publishers.

Bonner, P. and Segal, L. (1998) *Soweto: A History*, Cape Town: Maskew Miller Longman.

Browde, J., Mokhoba, P. and Jassat, E. (1998) *Report of the Internal Reconciliation Commission*, Johannesburg: University of the Witwatersrand, Faculty of Health Sciences.

Browett, J.G. (1976) 'The application of a spatial model to South Africa's development regions', *South African Geographical Journal* 58: 118–29.

Buchanan, A. (1991) *Secession: The Morality of Political Divorce from Fort Sumter to Lithuania and Quebec*, Boulder, CO: Westview.

Bundy, C. (1979) *The Rise and Fall of the South African Peasantry*, London: Heinemann.

Butler, J., Rotberg, R.I. and Adams, J. (1977) *The Black Homelands of South Africa: The Political and Economic Development of Bophuthatswana and KwaZulu*, Berkeley: University of California Press.

Cape of Good Hope (1905) *Census 1904*, G19-'05, Cape Town: Government Printer.

Christopher, A.J. (1976) *Southern Africa: Studies in Historical Geography*, Folkestone: Dawson.

—— (1983) 'From Flint to Soweto: reflections on the colonial origins of the apartheid city', *Area* 15: 145–9.

—— (1987) 'Race and residence in colonial Port Elizabeth', *South African Geographical Journal* 69: 3–20.

—— (1988a) *The British Empire at its Zenith*, London: Croom Helm.

—— (1988b) 'Roots of urban segregation: South Africa at Union 1910', *Journal of Historical Geography* 14: 151–69.

—— (1989) 'Apartheid within apartheid: an assessment of official intra-Black segregation on the Witwatersrand, South Africa', *Professional Geographer* 41: 328–36.

—— (1990) 'Apartheid and urban segregation levels in South Africa', *Urban Studies* 27: 421–40.

—— (1991) 'Changing patterns of group area proclamations in South Africa 1950–1989', *Political Geography Quarterly* 10: 240–53.

—— (1992) 'Urban segregation levels in the British Overseas Empire and its successors in the twentieth century', *Transactions of the Institute of British Geographers* 17: 95–107.

—— (1995a) 'Segregation and cemeteries in Port Elizabeth, South Africa', *Geographical Journal* 161: 38–46.

—— (1995b) 'Regionalism and ethnicity in South Africa 1990–1994', *Area* 28: 1–11.

—— (1995c) 'Land restitution in South Africa 1991–94', *Land Use Policy* 12: 267–79.

—— (1996) 'Regional patterns in South Africa's post-apartheid election 1994', *Environment and Planning C: Government and Policy* 14: 55–69.

—— (1997) *Spatial Aspects of Indigenous Lands and Land Claims in the Anglophone World*, Cambridge: University of Cambridge, Department of Land Economy.

—— (1999) *The Atlas of States: Global Change 1900–2000*, Chichester: John Wiley.

Cobbett, W. and Cohen, K. (1988) *Popular Struggles in South Africa*, London: James Currey.

Cock, J. and Nathan, L. (1989) *War and Society: The Militarisation of South Africa*, Cape Town: David Philip.

Coetzee, D. (1994) 'Vlakplaas and the murder of Griffiths Mxenge', in A. Minnaar, I. Liebenburg and C. Schutte (eds) *The Hidden Hand: Covert Operations in South Africa*, Pretoria: Human Sciences Research Council.

Coleman, M. (1998) *A Crime Against Humanity: Analysing the Repression of the Apartheid State*, Cape Town: David Philip.

Cook, G.P. (1986) 'Khayelitsha – policy change or crisis response?', *Transactions of the Institute of British Geographers* 11: 57–66.

Crankshaw, O. (1997) *Race, Class and the Changing Division of Labour under Apartheid*, London: Routledge.

Crankshaw, O. and Hart, T. (1990) 'The roots of homelessness: causes of squatting in Vlakfontein settlement south of Johannesburg', *South African Geographical Journal* 72: 65–70.

Crush, J.S. (1991) 'The chains of migrancy and the Southern African Labour Commission', in C. Dixon and M.J. Heffernan (eds) *Colonialism and Development in the Contemporary World*, London: Mansell.

—— (1999) 'Fortress South Africa and the deconstruction of apartheid's migration regime', *Geoforum* 30: 1–11.

Crush, J.S. and James, W. (1995) *Crossing Boundaries: Mine Migrancy in a Democratic South Africa*, Cape Town: Institute for Democracy in South Africa.

Cumming, S.D. (1993) 'Post-colonial urban residential change in Harare', in L.M. Zinyama, D.S. Tevera and S.D. Cumming (eds) *Harare: The Growth and Problems of the City*, Harare: University of Zimbabwe Publications.

da Costa, Y. and Davids, A. (1994) *Pages from Cape Muslim History*, Pietermaritzburg: Shuter and Shooter.

Davenport, T.R.H. (1970) 'The triumph of Colonel Stallard: the transformation of the Natives (Urban Areas) Act between 1923 and 1937', *South African Historical Journal* 2: 77–96.
—— (1971) *The Beginnings of Urban Segregation in South Africa: The Natives (Urban Areas) Act of 1923 and its Background*, Grahamstown: Rhodes University, Institute for Social and Economic Research
—— (1991) *South Africa: A Modern History*, London: Macmillan.
—— (1998) *The Transfer of Power in South Africa*, Cape Town: David Philip.
Davies, R.J. (1963) 'The growth of the Durban Metropolitan Area', *South African Geographical Journal* 45: 15–43.
—— (1981) 'The spatial formation of the South African city', *GeoJournal* (Supplementary Issue) 2: 59–72.
—— (1996) *Contemporary City Restructuring*, Johannesburg: Society of South African Geographers.
Davies, R.J. and Rajah, D.S. (1965) 'The Durban C.B.D.: boundary delimitation and racial dualism', *South African Geographical Journal* 47: 45–58.
Davies, W.J. (1971) *Patterns of Non-White Population Distribution in Port Elizabeth with Special Reference to the Application of the Group Areas Act*, Port Elizabeth: University of Port Elizabeth.
Davis, S.M. (1987) *Apartheid's Rebels: Inside South Africa's Hidden War*, New Haven, CT: Yale University Press.
de Beer, J. and Lourens, L. (1995) *Local Government: The Road to Democracy*, Midrand: Educum.
de Swardt, W.R.P. (1970) 'Urbanism and Bantu – focus on town planning and development in Soweto', in H.L. Watts (ed.) *Focus on Cities*, Durban: University of Natal, Institute of Social Research.
de Villiers, J. (1981) *A Regional Development Strategy for Southern Africa*, Pretoria: Department of Foreign Affairs.
Delport, A. (1997) *Official South African Local Government Yearbook 1997/8*, Johannesburg: Gaffney.
Department of Foreign Affairs List (1990–99) Pretoria: Department of Foreign Affairs.
Desmond, C. (1970) *The Discarded People: An Account of African Resettlement in South Africa*, Harmondsworth: Penguin.
Dick, H.W. and Rimmer, P.J. (1998) 'Beyond the Third World city: the new urban geography of South-east Asia', *Urban Studies* 35: 2303–21.
du Toit, B.M. (1991) 'The far right in current South African politics', *Journal of Modern African Studies* 29: 627–67.
Dubow, S. (1995) *Scientific Racism in Modern South Africa*, Cambridge: Cambridge University Press.
Duncan, O.D. and Duncan, B. (1955) 'A methodological analysis of segregation indexes', *American Sociological Review* 20: 210–17.
Durban (1923) *Mayor's Minute*, Durban: Durban Corporation.
Eastern Province Herald (1948–99) Port Elizabeth.
Elphick, R. and Giliomee, H. (1979) *The Shaping of South African Society (1652–1820)*, Cape Town: Longman.
Esterhuysen, P. (1982) 'Greater Swaziland?', *Africa Insight* 12: 181–8.
Evans, I.T. (1997) *Bureaucracy and Race: Native Administration in South Africa*, Berkeley: University of California Press.
Fair, T.J.D. (1982) *South Africa: Spatial Frameworks for Development*, Cape Town: Juta.
Geldenhuys, D. (1990) *Isolated States: A Comparative Analysis*, Johannesburg: Jonathan Ball.
Gibb, R.A. (1987) 'The effect on the countries of SADCC of economic sanctions against the Republic of South Africa', *Transactions of the Institute of British Geographers* 12: 398–412.

—— (1988) 'Southern Africa in transition: prospects and problems facing regional integration', *Journal of Modern African Studies* 30: 287–306.

—— (1998) 'Flexible integration in the "new" South Africa', *South African Geographical Journal* 80: 43–51.

Gibbon, A.D. (1977) 'Outdoor recreation survey of the Port Elizabeth area', unpublished MA thesis, University of Port Elizabeth.

Gigaba, M. and Maharaj, B. (1996) 'Land invasions during political transition: the Wiggins saga in Cato Manor', *Development Southern Africa* 13: 215–35.

Government Gazette (1948–91) Pretoria.

Griffiths, I. (1989) 'Airways sanctions against South Africa', *Area* 21: 249–59.

Griffiths, I. and Funnell, D.C. (1991) 'The abortive Swaziland deal', *African Affairs* 90: 51–64.

Gudgin, G. and Taylor, P.J. (1979) *Seats, Votes and the Spatial Organisation of Elections*, London: Pion.

Guelke, L. and Shell, R. (1983) 'An early colonial landed gentry: land and wealth in the Cape Colony 1682–1731', *Journal of Historical Geography* 9: 265–86.

Guillaume, P. (1997) 'Du blanc au noir . . . Essai sur une nouvelle ségrégation dans le centre de Johannesburg', *L'Espace Géographique* 26: 21–34.

Gutteridge, W. and Spence, J.E. (1997) *Violence in Southern Africa*, London: Frank Cass.

Hansard (1910–99) *Debates of Parliament*, Cape Town: Government Printer.

Hart, G. (1989) 'On grey areas', *South African Geographical Journal* 71: 81–93.

Hart, T. (1976) 'Patterns of Black residence in the White residential areas of Johannesburg', *South African Geographical Journal* 58: 141–50.

Heard, K.A. (1974) *General Elections in South Africa 1943–1970*, London: Oxford University Press.

Herbstein, D. and Evenson, J. (1989) *The Devils Are Among Us: The War for Namibia*, London: Zed Books.

Herholdt, A.D. (1986) *Die Argitektuur van Steytlerville met Rigline vir Bewaring en Ontwikkleling*, Port Elizabeth: Institute of South African Architects.

Horn, A.C. (1998) 'The identity of land in the Pretoria District, 19 June 1913: implications for land restitution', *South African Geographical Journal* 80: 9–22.

Huttenback, R.A. (1976) *Racism and Empire: White Settlers and Colored Immigrants in the British Self-Governing Colonies 1830–1910*, Ithaca, NY: Cornell University Press.

Hyam, R. (1972) *The Failure of South African Expansion*, London: Macmillan.

Inskeep, R.R. (1978) *The Peopling of Southern Africa*, Cape Town: David Philip.

Johnson, R.W. and Schlemmer, L. (1996) *Launching Democracy in South Africa: The First Open Election April 1994*, New Haven, CT: Yale University Press.

Judin, H. and Vladislavic, I. (1998) *blank——: Architecture, Apartheid and After*, Cape Town: David Philip.

Jurgens, U. (1993) 'Mixed race residential areas in South African cities – urban geographical developments in the late and post-apartheid phases', *GeoJournal* 30: 309–16.

Khosa, M.M. (1998) 'Employment and asset creation through public works in KwaZulu-Natal', *South African Geographical Journal* 80: 33–40.

Klotz, A. (1995) *Norms in International Relations: The Struggle against Apartheid*, Ithaca, NY: Cornell University Press.

Kotze, N.J. and Donaldson, S.E. (1998) 'Residential desegregation in two South African cities: a comparative study of Bloemfontein and Pietersburg', *Urban Studies* 35: 467–77.

Krige, D.S. (1996) *Botshabelo: Former Fastest Growing Urban Area in South Africa Approaching Zero Population Growth*, Bloemfontein: University of the Orange Free State, Department of Urban and Regional Planning.

—— (1997) 'Post-apartheid development challenges in small towns of the Free State', *South African Geographical Journal* 79: 175–8.

—— (1998) 'The challenge of dismantling spatial patterns constructed by apartheid in the Bloemfontein-Botshabelo-Thaba Nchu region', *Acta Academica* Supplementum 1: 174–218.

Kuper, L., Watts, H. and Davies, R.J. (1958) *Durban: A Study in Racial Ecology*, London: Jonathan Cape.

Lamar, H. and Thompson, L. (1981) *The Frontier in History: North America and Southern Africa Compared*, New Haven, CT: Yale University Press.

Land Info (1994–9) Pretoria: Department of Land Affairs.

Lehmann, S.G. and Ruble, B.A. (1997) 'From "Soviet" to "European" Yaroslavl: changing neighbourhood structure in post-Soviet Russian cities', *Urban Studies* 34: 1085–107.

Leistner, G.M.E. (1987) 'Africa at a glance', *Africa Insight* 17: 1–104.

Lemon, A. (1976) *Apartheid: A Geography of Separation*, Farnborough: Saxon House.

—— (1984) 'The Indian and coloured elections: co-option rejected?', *South Africa International* 15: 84–107.

—— (1987) *Apartheid in Transition*, Aldershot: Gower.

—— (1990) 'Geographical issues and outcomes in South Africa's 1989 "general" election', *South African Geographical Journal* 72: 54–64.

—— (1991a) *Homes Apart: South Africa's Segregated Cities*, London: Paul Chapman.

—— (1991b) 'Apartheid as foreign policy: the dimensions of international conflict in Southern Africa', in N. Kliot and S. Waterman (eds) *The Political Geography of Conflict and Peace*, London: Belhaven.

—— (1992) 'Restructuring the local state in South Africa: regional services councils, redistribution and legitimacy', in D. Drakakis-Smith (ed.) *Urban and Regional Change in Southern Africa*, London: Routledge.

—— (1995) *The Geography of Change in South Africa*, Chichester: John Wiley.

—— (1996) 'The new political geography of the local state in South Africa', *Malaysian Journal of Tropical Geography* 27(2): 35–45.

Maasdorp, G. (1980) 'Forms of partition', in R.I. Rotberg and J. Barratt (eds) *Conflict and Compromise in South Africa*, Cape Town: David Philip.

Mabin, A. (1986) 'Labour, capital, class struggle and the origins of residential segregation in Kimberley 1880–1920', *Journal of Historical Geography* 12: 4–26.

—— (1992) 'Comprehensive segregation: the origins of the Group Areas Act and its planning apparatuses', *Journal of Southern African Studies* 18: 405–29.

—— (1995) 'On the problems and prospects of overcoming segregation and fragmentation in South Africa's cities in a postmodern era', in S. Watson and K. Gibson (eds) *Postmodern Cities and Space*, Oxford: Oxford University Press.

Mabin, A. and Smit, D. (1997) 'Reconstructing South Africa's cities? The making of urban planning 1900–2000', *Planning Perspectives* 12: 193–223.

McKenzie, P. (1998) 'Reclaiming the land: a case study of Riemvasmaak', in J. Cock and P. McKenzie (eds) *From Defence to Development: Redirecting Military Resources in South Africa*, Cape Town: David Philip.

McKinley, D.T. (1997) *The ANC and the Liberation Struggle: A Critical Political Biography*, London: Pluto Press.

McNeil, K. (1989) *Common Law Aboriginal Title*, Oxford: Clarendon Press.

Maharaj, B. (1994) 'The Group Areas Act and community destruction in South Africa: the struggle for Cato Manor in Durban', *Urban Forum* 5(2): 1–25.

—— (1995) 'The local state and residential segregation: Durban and the prelude to the Group Areas Act', *South African Geographical Journal* 77: 33–41.

—— (1996a) 'The historical development of the apartheid local state in South Africa: the case of Durban', *International Journal of Urban and Regional Research* 20: 587–600.

—— (1996b) 'Urban struggles and the transformation of the apartheid local state: the case of community and civic organisations in Durban', *Political Geography* 15: 61–74.

—— (1997) 'The politics of local government restructuring and apartheid transformation in South Africa', *Journal of Contemporary African Studies* 15: 261–85.

—— (1999) 'The integrated community apartheid could not destroy: the Warwick Avenue Triangle in Durban', *Journal of Southern African Studies* 25: 249–66.

Maharaj, B. and Mpungosa, J. (1994) 'The erosion of residential segregation in South Africa: the "greying" of Albert Park, Durban', *Geoforum* 25: 19–32.

Malan, T. (1980) *Two Views on South Africa's Foreign Policy and the Constellation of States*, Johannesburg: South African Institute of Race Relations.

Malan, T. and Hattingh, P.S. (1976) *Black Homelands in South Africa*, Pretoria: Africa Institute.

Mamdani, M. (1996) *Citizen and Subject: Contemporary Africa and the Legacy of Late colonialism*, Princeton, NJ: Princeton University Press.

Massey, D.S. and Denton, N.A. (1989) 'Hypersegregation in U.S. metropolitan areas: Black and Hispanic segregation along five dimensions', *Demography* 26: 373–91.

—— (1993) *American Apartheid: Segregation and the Making of the Underclass*, Cambridge, MA: Harvard University Press.

Mather, C. (1987) 'Residential segregation and Johannesburg's "locations in the Sky"', *South African Geographical Journal* 69: 119–28.

Mattera, D. (1987) *Memory Is the Weapon*, Johannesburg: Ravan.

Mbeki, G. (1996) *Sunset at Midday*, Johannesburg: Nolwazi.

Mbeki, T. (1998) *Africa: The Time Has Come*, Cape Town: Tafelberg.

Meli, F. (1988) *South Africa Belongs to Us: A History of the ANC*, Harare: Zimbabwe Publishing House.

Monthly Abstract of Trade Statistics (1955–99) Pretoria: Government Printer.

Moolman, H.J. (1977) 'The creation of living space and homeland consolidation with reference to Bophuthatswana', *South African Journal of African Affairs* 7: 149–63.

Morris, A. (1994) 'The desegregation of Hillbrow, Johannesburg 1978–1982', *Urban Studies* 31: 821–34.

—— (1997) 'Physical decline in an inner-city neighbourhood: a case study of Hillbrow, Johannesburg', *Urban Forum* 8: 153–75.

Morris, P. (1980) *Soweto: A Review of Existing Conditions and Some Guidelines for Change*, Johannesburg: Urban Foundation.

Mostert, N. (1992) *Frontiers: The Epic of South Africa's Creation and the Tragedy of the Xhosa People*, London: Jonathan Cape.

Murray, C. (1992) *Black Mountain: Land, Class and Power in the Eastern Orange Free State 1880s–1980s*, Edinburgh: Edinburgh University Press.

Murray, C. and O'Regan, C. (eds) (1990) *No Place to Rest: Forced Removals and the Law in South Africa*, Cape Town: Oxford University Press.

Nel, J.G. (1988) *Die Geografiese Impak van die Wet op Groepsgebiede en Verwante Wetgewing op Port Elizabeth*, Port Elizabeth: University of Port Elizabeth.

Nitz, H.J. (1993) *The Early-Modern World-System in Geographical Perspective*, Stuttgart: Franz Steiner.

Oelosfse, C. and Dodson, B. (1997) 'Community, place and transformation: a perceptual analysis of residents' responses to an informal settlement in Hout Bay, South Africa', *Geoforum* 28: 91–101.

O'Meara, D. (1996) *Forty Lost Years: The Apartheid State and the Politics of the National Party 1948–1994*, Johannesburg: Ravan.

Oosterlig (1948–89) Port Elizabeth.

Pakenham, T. (1979) *The Boer War*, London: Weidenfeld and Nicolson.

Parnell, S. (1986) 'From Mafeking to Mafikeng: the transformation of a South African town', *GeoJournal* 12: 203–10.

—— (1988) 'Land acquisition and the changing residential face of Johannesburg', *Area* 20: 307–14.

—— (1989) 'Shaping a racially divided society: state housing policy in South Africa 1920–1950', *Environment and Planning C: Government and Policy* 7: 261–72.

—— (1991) 'Sanitation, segregation and the Natives (Urban Areas) Act: African exclusion from Johannesburg's Malay Location 1897–1925', *Journal of Historical Geography* 17: 271–88.

Payne, R. (1992) *The Two South Africas: A People's Geography*, Johannesburg: Human Rights Commission.

Peberdy, S. and Crush, J. (1998) 'Rooted in racism: the origins of the Aliens Control Act', in J. Crush (ed.) *Beyond Control: Immigration and Human Rights in a Democratic South Africa*, Cape Town: Southern African Migration Project.

Pelzer, A.N. (1966) *Verwoerd Speaks: Speeches 1948–1966*, Johannesburg: APB Publishers.

Pieres, J.B. (1995) 'Ethnicity and pseudo-ethnicity in the Ciskei', in W. Beinart and S. Dubow (eds) *Segregation and Apartheid in Twentieth-Century South Africa*, London: Routledge.

Pinnock, D. (1984) *The Brotherhoods: Street Gangs and State Control in Cape Town*, Cape Town: David Philip.

Pirie, G.H. (1984) 'Race zoning in South Africa: board, court, parliament, public', *Political Geography Quarterly* 3: 207–21.

—— (1990a) 'Racial segregation on public transport in South Africa 1877–1988', unpublished PhD thesis, University of the Witwatersrand.

—— (1990b) 'Aviation, apartheid and sanctions: air transport to and from South Africa, 1945–1989', *GeoJournal* 22: 231–40.

Pirie, G.H. and Hart, D. (1985) 'The transformation of Johannesburg's Black western areas', *Journal of Urban History* 11: 387–410.

Pirie, G.H. Rogerson, C.M. and Beavon, K.S.O. (1980) 'Covert power in South Africa: geography of the Afrikaner Broederbond', *Area* 12: 97–104.

Platzky, L. and Walker, C. (1985) *The Surplus People: Forced Removals in South Africa*, Johannesburg: Ravan.

Port Elizabeth (1999) *The First Comprehensive Urban Plan*, Port Elizabeth: Port Elizabeth Municipality.

Posel, D. (1991) *The Making of Apartheid 1948–1961: Conflict and Compromise*, Oxford: Clarendon Press.

Ramphal, S. (1986) *Mission to South Africa: The Commonwealth Report*, Harmondsworth: Penguin.

—— (1989) *South Africa: The Sanctions Report*, Harmondsworth: Penguin.

Ramutsindela, M.F. (1997) 'National identity in South Africa: the search for harmony', *GeoJournal* 43: 99–110.

—— (1998) 'The changing meanings of South Africa's internal boundaries', *Area* 30: 291–9.

Reynolds, A. (1994) *Election '94 South Africa: The Campaigns, Results and Future Prospects*, London: James Currey.

Ritchten, E. (1989) 'The KwaNdebele struggle against independence', *South African Review* 5: 426–45.

Robinson, J. (1996) *The Power of Apartheid: State, Power and Space in South African Cities*, London: Butterworth-Heinemann.

Rogerson, C.M. (1990) 'Defending apartheid: Armscor and the geography of military production in South Africa', *GeoJournal* 22: 241–50.

—— (1997) 'African immigrant entrepreneurs and Johannesburg's changing inner city', *Africa Insight* 27: 265–73.

Rule, S.P. (1989) 'The emergence of a racially mixed residential suburb in Johannesburg: the demise of the apartheid city?', *Geographical Journal* 155: 196–203.

—— (1994) 'A second-phase diaspora: South African migration to Australia', *Geoforum* 25: 33–9.

Saff, G. (1996) 'Claiming a space in a changing South Africa: The "squatters" of Marconi Beam, Cape Town', *Annals of the Association of American Geographers* 86: 235–55.

Saunders, C. (1992) *Writing History: South Africa's Urban Past and Other Essays*, Pretoria: Human Sciences Research Council.

—— (1994) *An Illustrated History of South Africa: The Real Story*, Cape Town: Reader's Digest.

Saunders, C. and Southey, N. (1998) *A Dictionary of South African History*, Cape Town: David Philip.

Seegers, A. (1996) *The Military and the Making of Modern South Africa*, London: Tauris.

Seethal, C. (1991) 'Restructuring the local state in South Africa: Regional Services Councils and crisis resolution', *Political Geography Quarterly* 10: 8–25.

Shirer, W.L. (1964) *The Rise and Fall of the Third Reich*, London: Pan.

Simon, D. (1986) 'Desegregation in Namibia: the demise of urban apartheid?', *Geoforum* 17: 289–307.

—— (1989) 'Crisis and change in South Africa: implications for the apartheid city', *Transactions of the Institute of British Geographers* 14: 189–206.

—— (1995) *Strategic Territory and Territorial Strategy: The Geopolitics of Walvis Bay's Reintegration into Namibia*, Windhoek: Namibian Economic Policy Research Unit.

Simpson, M. (1993) 'Nation-building: a post-apartheid super glue?', *Indicator South Africa* 10(3): 17–20.

Smith, D.M. (ed.) (1992) *The Apartheid City and Beyond: Urbanization and Social Change in South Africa*, London: Routledge.

Smith, K. (1988) *The Changing Past: Trends in South African Historical Writing*, Johannesburg: Southern Books.

South Africa (1918–97) *Year Book* (various titles), Pretoria: Government Printer.

—— (1943) *Report of the Second Indian Penetration (Durban) Commission*, UG21-1943, Pretoria: Government Printer.

—— (1948) *Report of the Native Laws Commission 1946–1948*, UG28-1948, Pretoria: Government Printer.

—— (1950) *Statutes of the Union of South Africa 1950*, Parow: Government Printer.

—— (1954) *Population Census 1946*, UG18-1954, Pretoria: Government Printer.

—— (1955a) *Population Census 1951*, UG42-1955, Pretoria: Government Printer.

—— (1955b) *Summary of the Report of the Commission for the Socio-Economic Development of the Bantu Areas within the Union of South Africa*, UG61-1955 (Tomlinson Report), Pretoria: Government Printer.

—— (1960) *Report of the Commission of Inquiry into the European Occupancy of the Rural Areas*, Pretoria: Government Printer.

—— (1963) *Population Census 1960*, RP62-1963, Pretoria: Government Printer.

—— (1964) *Report of the Commission of Enquiry into South West African Affairs 1962–63*, RP12-1964, Pretoria: Government Printer.

—— (1973) *Population Census 1970*, Report No. 02-05-01, Pretoria: Government Printer.

—— (1980) *Report of the Inquiry into the Riots at Soweto and Elsewhere*, RP55-1980, Pretoria: Government Printer.

—— (1986) *Population Census 1985*, Report No. 02-85-01, Pretoria: Government Printer.

—— (1995) *Annual Report of the Department of Land Affairs 1994*, RP102-1995, Pretoria: Government Printer.

—— (1997) *Self-determination for Afrikaners*, RP105-1997, Pretoria: Government Printer.

—— (1998a) *Defence in a Democracy: South African Defence Review 1998*, Pretoria: Government Printer.

—— (1998b) *Population Census 1996*, Report No. 03-01-11, Pretoria: Statistics South Africa.

—— (1998c) *The Truth and Reconciliation Commission of South Africa, Report*, Cape Town: Truth and Reconciliation Commission.

—— (1999) *Annual Report of the Department of Land Affairs 1998*, RP41-1999, Pretoria: Government Printer.

South African Survey (1996–9) Johannesburg: Institute of Race Relations.

Spink, K. (1991) *Black Sash: The Beginning of a Bridge in South Africa*, London: Methuen.

Stalker, P. (1999) *Working Without Frontiers: The Impact of Globalization on International Migration*, Boulder, CO: Lynne Rienner.

Statesman's Year Book (1979–87) London: Macmillan.

Stern, E. (1987) 'Competition and the location of the gaming industry: the "casino states" of Southern Africa', *Geography* 72: 140–50.

Stiff, P. (1999) *The Silent War: South African Recce Operations 1969–1994*, Alberton: Galago.

Study Commission on U.S. Policy Toward Southern Africa (1981) *South Africa: Time Running Out*, Berkeley: University of California Press (cited as Study Commission).

Sunday Times (1989–99) Johannesburg.

Survey of Race Relations in South Africa (1951–95) Johannesburg: South African Institute of Race Relations.

Swan, M. (1985) *Gandhi: The South African Experience*, Johannesburg: Ravan.

Swanson, M.W. (1976) ' "The Durban system": the roots of urban apartheid in colonial Natal', *African Studies* 35: 159–76.

—— (1977) 'The sanitation syndrome: bubonic plague and urban native policy in the Cape Colony 1900–1909', *Journal of African History* 18: 387–410.

Tatz, C.M. (1962) *Shadow and Substance in South Africa: A Study in Land and Franchise Policies Affecting Africans 1910–1960*, Pietermaritzburg: University of Natal Press.

Thomas, S. (1996) *The Diplomacy of Liberation: The Foreign Relations of the African National Congress since 1960*, London: Tauris Academic Studies.

Thompson, L. (1990) *A History of South Africa*, New Haven, CT: Yale University Press.

Time (1977–99) New York.

Torr, L. (1987) 'Providing for the "better-class native", the creation of Lamontville', *South African Geographical Journal* 69: 31–46.

Transvaal (1922) *Report of the Transvaal Local Government Commission*, TP1-1922, Pretoria: Government Printer.

Turner, J.W. (1998) *Continent Ablaze: The Insurgency Wars in Africa 1960 to the Present*, Johannesburg: Jonathan Ball.

United Nations (1948–97) *Statistical Yearbook*, New York: United Nations.

Unterhalter, E. (1987) *Forced Removals: The Division, Segregation and Control of the People of South Africa*. London: International Defence and Aid Fund.

Vail, L. (1991) *The Creation of Tribalism in Southern Africa*, Berkeley: University of California Press.

van Onselen, C. (1982) *Studies in the Social and Economic History of the Witwatersrand 1886–1914: New Babylon, New Nineveh*, Johannesburg: Ravan.

—— (1996) *The Seed Is Mine: The Life of Kas Maine, a South African Sharecropper 1894–1985*, Cape Town: David Philip.

Weekly Mail & Guardian (1992–9) Johannesburg.

Wellington, J.H. (1967) *South West Africa and Its Human Issues*, Oxford: Clarendon Press.

Welsh, F. (1998) *A History of South Africa*, London: HarperCollins.

Western, J. (1996) *Outcast Cape Town*, Berkeley: University of California Press.

Wilkins, I. and Strydom, H. (1978) *The Super Afrikaners*, Johannesburg: Jonathan Ball.

Wilson, M. and Thompson, L. (1969) *The Oxford History of South Africa*, Vol. 1: *South Africa to 1870*, Oxford: Clarendon Press.

Worden, N. (1995) *The Making of Modern South Africa: Conquest, Segregation and Apartheid*, Oxford: Blackwell.

Yap, M. and Man, D.L. (1996) *Colour, Confusion and Concessions: The History of the Chinese in South Africa*, Hong Kong: Hong Kong University Press.

Index